the body multiple

Science and Cultural Theory   *A Series Edited by Barbara Herrnstein Smith and E. Roy Weintraub*

duke university press   durham and london   2002

annemarie mol

the body multiple ontology in medical practice

© 2002 Duke University Press

All rights reserved. Printed in the United

States of America on acid-free paper ⊗ Designed by

Amy Ruth Buchanan. Typeset in Scala and Scala Sans

by Tseng Information Systems, Inc. Library of Congress

Cataloging-in-Publication Data appear on the last

printed page of this book.

contents

Preface   vii

1. Doing Disease   1

2. Different Atheroscleroses   29

3. Coordination   53

4. Distribution   87

5. Inclusion   119

6. Doing Theory   151

Bibliography   185

Index   191

This is a book about the way in which (Western, cosmopolitan, allopathic) medicine deals with the body and its diseases. The questions it raises do not concern
the ways in which medicine *knows* its objects. Instead, what the book explores
is the ways in which medicine attunes to, interacts with, and shapes its objects
in its various and varied practices. Or, to use the technical term: this is a book
about the way medicine *enacts* the objects of its concern and treatment.

Thus, unlike many other books on medicine and its processes, this one does
not speak of different perspectives on the body and its diseases. Instead it tells
how they are done. This means that the book comes to talk about a series of
different practices. These are practices in which some entity is being sliced,
colored, probed, talked about, measured, counted, cut out, countered by walking, or prevented. Which entity? A slightly different one each time. Attending
to enactment rather than knowledge has an important effect: what we think of
as a single object may appear to be more than one. All the examples in this book
concern atherosclerosis. But a plaque cut out of an atherosclerotic artery is not
the same entity as the problem a patient with atherosclerosis talks about in the
consulting room, even though they are both called by the same name. The loss
of blood pressure over a stenosis is not the same thing as the loss of blood vessel
lumen that radiologists make visible on their X-ray pictures.

The move, then, is away from epistemology. Epistemology is concerned with
reference: it asks whether representations of reality are accurate. But what becomes important if we attend to the way objects are enacted in practices is quite
different. Since enactments come in the plural the crucial question to ask about

them is how they are coordinated. In practice the body and its diseases are more than one, but this does not mean that they are fragmented into being many. This is difficult to think. But it is this complex state of affairs that this book explores. I have tried to capture it in the title, in which a singular noun comes with a pluralizing adjective. This, then, is a book about an intricately coordinated crowd: *the body multiple*.

The tone of the text is reflective rather than argumentative. I have no reason either to criticize or to defend medicine as a whole—as if it *were* a whole. Instead of creating a position outside medicine in order to judge it, I try to engage with a normativity of a more intimate kind. I try to open up differences inside medicine and create better access to them. If the objects of medicine are enacted in a variety of ways, truthfulness is no longer good enough. Somehow, questions need to be asked about the appropriateness of various enactments of the body multiple and its diseases. I don't ask such questions here. I don't delve into the question of how the appropriateness of the various enactments presented are, or might be, judged. Instead I try to take part in creating a theoretical repertoire for thinking about this. I contribute to theorizing medicine's *ontological politics*: a politics that has to do with the way in which problems are framed, bodies are shaped, and lives are pushed and pulled into one shape or another.

Its concern with theorizing turns this into a philosophical book. But the philosophy I engage in here is of a quite specific kind. It is explicit about its local origins. Thus, throughout the book there are snapshot-stories about a single multiple disease and the way it is dealt with in a single hospital and some of its surroundings. The disease is atherosclerosis, and more particularly atherosclerosis of the leg arteries. The hospital is a large university hospital in a medium-sized Dutch town, anonymized into hospital Z. By starting out from such a well-circumscribed site, I try to move philosophy away from formats that carry universalistic pretentions, but that in fact hide the locality to which they pertain. However, the idea is not to celebrate localism instead of universalism. Instead, it is to keep track as persistently as possible of what it is that alters when matters, terms, and aims travel from one place to another.

Medical anthropology and medical sociology are rich disciplines. Thus, I had a lot to build on as I sought to incorporate an empirical investigation into my philosophical study. So much so that I have framed this book not only as a debate with the epistemological approach to knowledge, but also as a debate with the way in which the social sciences have studied the body and its diseases in the past. For a long time, social scientists have said that there is *more* than the physicalities treated by doctors. And then they used to study this "more": a social and an interpretative reality. They have differentiated between *disease* and *illness,* taking the latter as their object of study. More recently, the medical perspective on disease has been included in the studies, too. This book is among those who try to take the next move. It says that a study of the enactment of reality in practice makes it possible to ethnographically explore the body multiple and its diseases in all their fleshiness. How? Outlining an answer to that question is precisely what all these pages are for.

The book draws on a variety of literatures: in philosophy, anthropology, science and technology studies, feminist theory, sociology, political theory. This is the present state of theoretical work: disciplinary boundaries get blurred. And yet I wanted to give you, the reader, a good sense of where this book is situated. I wanted to ground it not only in empirical "material," but also in the intellectual traditions of which it is a product. After hesitating for quite a while about how to do this, I have turned this question into a topic. Throughout this book you will find a subtext, in which I relate to the literature (or, more exactly, to exemplary books and articles) while self-reflexively wondering what it is to do so.

Readers who regularly surf between television channels will find this book easier to read than those who don't, since they are likely to find out how to shift between the upper text and the subtext more quickly. Others will have to invent a way of reading that works for them from scratch. It may help to know that the subtext is not glued to the pages where it happens to be printed—its location is even more contingent than that of footnotes tends to be. Depending on where and who and how you are, you may want to read the subtext before you read a chapter, or afterward, or maybe when the story line of the upper text starts to bore you and you are in the mood for something different. It is up to you.

The book is written in English. This hides the plurality of the languages that went into its production. In the literature I draw on a few texts in German and a very small number in Dutch (although I have learned a great deal from reading around in my mother tongue). A large part of the literature I relate to was written and read in French. A lot more was in English. As part of my fieldwork I attended some English-language medical conferences and read English-language medical textbooks and research articles (some of them written by my local Dutch informants). But during the day-to-day events in the hospital the language spoken was almost always Dutch. And I also made my field notes in this language. Discussions about the many earlier versions of (parts of) this text were conducted in English, French and again, mostly Dutch.

Thus, though Dutch was a relevant language in the production of this book, in its final version it has vanished. What to say about this? Dutch is understood by only some 25 million people in a few regions of the world (mainly in the Netherlands, Surinam, Belgium, and South Africa where some of those speaking Afrikaans manage to comprehend Dutch—in Indonesia people with a good command of the language of their former colonizer are getting more rare every day). The Dutch failed to combine economic and cultural imperialism, so these days Dutch doesn't travel far. This means that a Dutch language intellectual must make a choice between being *local* or *global*. This choice has little to do with seeking a *small* or a *large* audience. Even if there are far more than 25 million people able to read English, most scholarly texts printed in Dutch are printed in more or less the same numbers as similar books in English. The local is not contained in the global. It is somewhere else.

Its language, then, marks this book as an academic text, made to travel through universities, to be read by scholars and students. I regret it that an attempt to reach my "international" colleagues obliges me to write in a foreign tongue, for that not only brings a lot of extra hard work, but also helps to widen the gap between embodied and inscribed author. Although a book I would publish in Dutch would be read by academic colleagues in neighboring fields as well as by many a Dutch physician, most of these possible readers are far less likely to come across this one. But then again: I am also deeply pleased to not be stuck in Dutchness, but to have been given a chance to acquire access to a language that allows one to reach readers from Norway to India, from anthropology to philosophy, from Germany to Brazil, from medicine to sociology, from the United States to France, and from science and technology studies to feminist theory. Or sometimes texts do not travel at all. That, again, is up to you, reader.

And now for some private history, as introductions go.

The fieldwork for this text started in the early seventies, when, over dinner at the kitchen table, my father told me about his work on using Doppler measurements for the assessment of the carotid arteries. From long before I officially interviewed him, he has been a wonderful informant. My mother engaged with the second feminist wave in the late sixties, turning me into a feminist at age eleven. As a geographer she also made me attentive to the spatiality of landscapes, townscapes, and life in general, for which I thank her.

But this book only really got under way in the academic year 1977–78. I was in the second year of medical school by then and a first-year student of philosophy. Thursdays were the best. In the mornings I had a philosophy class about the body and in the afternoons an anatomy class where we dissected corpses. Barthes gave way to a large, white room that stank of formalin. Merleau-Ponty was followed by corpses wrapped in orange towels and green plastic. In the mornings I would learn to unravel Foucault's writings and in the afternoons I was supposed to explore the pelvic cavity of a female body without cutting through nerves and blood vessels. This is more than twenty years ago and yet this book is to some extent a product of those long-gone Thursdays, not in the least the remarkable materiality of it all: sentences in difficult French, strange smells, my clumsiness in cutting.

For their help in the intermediate years I would like to thank various people. First of all Peter van Lieshout, with whom I wrote about the coexistence of "ontologies" in the early eighties and later about social theory and the delineation of the object of care in Dutch general practice and mental health care. All along he also helped me to tame the complexity of life—even if he increased it too, if only by fathering Elisabeth and Johannes, our children, whom I thank for being. Jan van Es made it possible for me to become a theorist of medicine in medical school. Lolle Nauta and Gerard de Vries tried to teach me how to argue. Dick Willems shared his energy and his investigations into medicine with me. Jeannette Pols worked on this project with remarkable zeal. Marc Berg and Ruud Hendriks did great work as well, in their shifting roles of research assistant, co-author, and coeditor. Agnes Vincenot, Pieter Pekelharing, Jan Willem Duyvendak, Sigrid Sijthoff, Tsjalling Swierstra, Bernike Pasveer, Hans Harbers, Marja Gastelaars, Sjaak Koenis, Rob Hagendijk, Rein de Wilde, Cor van der Weele, Eddy Houwaart, Baukje Prins, Paul Wouters, Evelien Tonkens, Marianne van den Boomen, Berteke Waaldijk, Mieke Aerts, Jens Lachmund, and Geertje Mak gave support both intellectually and otherwise. I have also learned a great deal from working with Bernard Elsman, Ant Lettinga, Bart van Lange, Antoinette de Bont, Jessica Mesman, Ineke Klinge, Ariane de Ranitz, Brenda Diergaarde, Irma van der Ploeg, Amâde M'charek, Tiago Moreira, Benedicte Rousseau, Alice Stollmeijer, and Toine Pieters in various modes and modalities. I would like to thank Barbara Duden, Donna Haraway, and Marilyn Strathern for the example they set and the work they do and Bruno Latour and Michel Callon for their challenge and encouragement. It was good to sometimes come across Sarah Franklin, Isabelle Baszanger, Charis Thompson, Madeleine Akrich, Vololona Rabeharisoa, Ingunn Moser, Claudia Castañeda, and Vicky Singleton and so feel that I was part of an international *current*. Nicolas Dodier asked the right questions at the right time, and Stefan Hirschauer incited me to be ever more serious. Marianne de Laet listened to my stories and gave careful comments on a previous version. Three reviewers of Duke University Press, whose names I do not know, finally approved of this manuscript, but before that came up with a lot of valuable, constructive criticism. And so did Noortje Marres, who figured as a fresh reader for the penultimate version. John Law attended time and again to all the details of this book, improved on many of them, invented new rhizomes, coauthored and wrote about related topics, corrected the English of several consecutive versions, and pushed me to come to a conclusion. That is a lot. Thanks. To you all.

And finally I would like to thank my informants. Here I begin with Ab Struyvenberg, who welcomed me into the hospital just before he retired and who kept reading my drafts afterward. Of course I could not have done any fieldwork at all without the collaboration of the many doctors, nurses, technicians, researchers, and patients of hospital Z who allowed me to observe and question them. They not only gave me material to think about and to think with, but in some cases also commented on my writings. Going along with the ethnographic habit of protecting the identities of informants, I mention no names here. But I am all the more grateful for their time and their trust.

For its generous financial support I thank the Netherlands Organization for Scientific Research, which provided me with a Constantijn and Christiaan Huygens grant that allowed me to do research and write about *Differences in Medicine* for five years. Later grants, notably of the ethics and policy section of this same organization, allowed me to continue to write on new topics and themes, meanwhile spending some of my time on revisions of and corrections to this book.

Even if in the end I wrote alone, I don't particularly want to be blamed for the remaining errors. I would, instead, be very grateful to you, reader, if you were to point them out and improve on them in your own writings.

A Movement between Fields

This is a study in empirical philosophy. Let's begin with the empirical. The stories I will tell you in this book are mostly situated in a university hospital in a medium-sized town in the center of the Netherlands, *Hospital Z*. For four years I went there once or twice weekly. I had an identity card that allowed me to leave my bicycle behind a fence and drink free coffee from the omnipresent vending machines. I had a library card and the use of a desk in a succession of crowded rooms. I had a white coat. And I observed.

I would go to the professor who headed a department and explain my purpose: to investigate the way the tensions between sources of knowledge and styles of knowing are handled inside present-day allopathic medicine — or at least one of its exemplars. I would explain what made "atherosclerosis in the lower limbs" a suitable case for my purpose and what I hoped to learn in their department. I presented myself as both insider and outsider, having received basic training in medical school as well as extensive training in philosophy. And I gave the name of the professor of internal medicine supporting my study. Each of the professors thus approached reacted in a friendly way. They all emphasized that academic hospitals must encourage research. My particular research plans made some interested and some skeptical. Others simply were indifferent. But after some further questions I would invariably be sent to someone one or more steps down the hierarchy to talk about and practically arrange my observation.

So I sat for many mornings behind vascular surgeons and internists doing their outpatient clinics, observing some three hundred consultations. (All surgeons and internists I observed for this study were men, and I will not hide that fact, so I use the generic "he" whenever I write about "the doctor," even though one of the pathologists whom I observed was a woman. Yes, this is a fading historical moment. The profession is undergoing a rapid gender change. But that is another story. One more complication left out here.) In university hospitals, both physicians and patients are used to observers: there are always students and junior doctors around who need to learn something. Yet I was surprised by the calm with which my presence was accepted—for I found these observations rather intimate. Patients tell about so much and undress so often. Although that is difficult for some and a relief to others, my presence behind the attending doctor hardly seemed to make a difference. When it risked to do so, I skipped a visit (once when a patient asked for it, several times when a doctor did, and once when I recognized someone I knew vaguely and left of my own initiative). The other transgression was into the privacy of the doctors. I was in a position to observe all kinds of details about the way they work. Some of them were visibly uneasy about the fact that I might judge the degree to which they were humane and kind in their interactions with patients. But (though that was sometimes difficult to resist) I wasn't out to make such judgments. Nor did I want to judge the so-called technicalities of their diagnosis and treatment. I wanted my obser-

How to Relate to the Literature?

In the ethnographic stories that I tell throughout this book, I do not try to sum things up. I do not describe Western medicine, but particular events in a single Dutch university hospital. And I assume that events in the next hospital, thirteen kilometers away, or over the border in Germany, or across the Atlantic have a complex relation with those that I have witnessed. A comparative analysis would show that there are similar patterns. Similar gestures. Similar machines. But also different self-evidences. Different needles and different norms. Different jokes. But which differences exactly? And what are their interferences and their diffractions? I haven't studied this. The relations of similarity and difference between one medical site and another are a topic in their own

right. By leaving that topic open I at least avoid the risk of answering it in the standard way. I avoid assuming that what happens in a single hospital forms part of a larger system of medicine: Western, cosmopolitan, modern, allopathic. If one assumes the existence of such a system, one can then be unpleasantly surprised by the amount of "medical practice variation."

But where is the standard way of understanding medicine as a system to be found? And where are the surprises that come with finding "variations"? Not exactly in the hospital I studied, where these things are hardly a matter for debate. No. They are to be found in the literature (see, e.g., Andersen and Mooney 1990). So what I have to tell in the present book does not just relate to the events that figure in my stories. It also relates to other texts. Lots

vations to be a means to get to know their standards, rather than an occasion to apply my own.

This made me shift sites and move around in the hospital. I observed technicians handling diagnostic tools in the vascular laboratory. I followed the tracks of radiologists and pathologists in their dealings with leg arteries. I went for months to the weekly meetings where the treatment options for patients with complicated cases of vascular disease were discussed. I witnessed several operations. Spent some days in the research laboratory of the hematologists. Held interviews or had conversations with epidemiologists, physiologists, internists, surgeons, and general practitioners. A couple of them read my articles and we talked about their reactions. I also went to the library and studied the textbooks and journal articles written, or mobilized as a resource, by "my doctors" and, when the references and my curiosity took me there, compared them with other publications. For two years I followed the monthly research colloquium on atherosclerosis. I coauthored with a junior doctor an article about the introduction of a diagnostic protocol. I supervised a medical student who interviewed vascular surgeons in several smaller hospitals and another one who analyzed discussions about the intake of cholesterol. And, finally, I had the temporary luxury of a research assistant—Jeannette Pols, a philosopher like myself, moreover trained as a psychologist—who held long patient interviews, transcribed them, talked them over with me, and coauthored publications about this material. She also was a good sparring partner with whom to discuss my work.

of them. Texts about other hospitals and other medical practices, texts about bodies and diseases, but also texts about entirely different topics. Systems and events, controversies, similarities and differences, coexistence, methods, politics. If I am to make explicit how *this* text departs from the others around it, if I want to show how it both differs from them *and* is made possible by them, I will have to *relate to the literature*. But how to do this? How to relate to the literature? That is a question that I take very seriously. So I have not hidden the answer between the lines. I do not follow one of the genres for using literatures without being explicit about it. Instead I have tried—will try—both to relate to the

literature and deal with the question as to how one might do so. To do this properly, I have separated out the question about relating to the literature from the core text of this book. I deal with the literature in a series of separate texts that resonate, run along, interfere with, alienate from, and give an extra dimension to the main text. In a subtext, so to speak.

Specificities
Relating to the literature, I might write: "In a variety of disciplines, the unity of Western medicine was a trope for decades. In medical sociology the unity of the medical profession explained this profession's social power. In medical anthropology the

Discussion was also what I sought in other worlds, outside the hospital. I could seldom go to those places by bicycle, for they were a lot farther away—and yet they were less alien to my writing and talking self. They were departments of philosophy, anthropology, sociology, or science and technology studies. I attended conferences and listened bored or fascinated to speakers presenting papers to five or fifty listeners. I read journal articles, wrote them, reviewed them. I went for talk-walks on lakesides or chatted over dinners. I was cross-examined about my field, my method, my purpose, my theoretical ancestors. Often such exchanges took place in an odd version of the English language, a transportation device that poses some difficulties to those who have not grown up with it, but reaches far. So though my stories come from the hospital in the town where I live, they went with me to many other places. To my intellectual friends and enemies in places like Maastricht, Bielefeld, Lancaster, Paris, Montreal, San Francisco. They managed to travel, my stories about leg vessels and pain. Immersed in theoretical arguments about the multiplication of reality.

For even if there are a lot of empirical materials in this book, this is not a field report: it is an exercise in *empirical philosophy*. Let's shift to the philosophy. The plot of my stories about vessels and fluids, pain and technicians, patients and doctors, techniques and technologies in hospital Z is part of a philosophical narrative. In conformity with the dominant habit of that genre, I'll give away the plot right here, at the beginning. It is this. It is possible to refrain from understanding objects as the central points of focus of different people's perspectives. It is possible to understand them instead as things manipulated in practices. If

divergence of medical traditions from all over the globe was specified by contrasting these traditions with a solid unity called Western medicine (either in order to show the superstitious character of the Others, or to highlight their ingenuity and greater sensitivity). In medical history the old eclecticism in which many schools and skills coexisted was turned into an intriguing counterpoint to the present homogeneity. And medical philosophy took a unity, the person-as-a-whole, as a norm: its wholeness deserved respect." Indeed, I have written (or rather coauthored) something like that. Elsewhere. (For a slightly longer version of such an overview, see Mol and Berg 1998, 1–12.)

It is possible to relate to the literature in such a way: evoking four entire disciplines, in just a few lines. The level of generality is a bit overwhelming. So much so that it is hardly feasible to insert titles. Sure, this can be done. After each discipline a name and date may be put between brackets. In medical sociology in the seventies . . . (see, e.g., Freidson 1970). A gesture like that turns Freidson's *The Profession of Medicine* into a representative of the enormous pile of books and articles published in the 1970s under the heading "medical soci-

we do this—if instead of bracketing the practices in which objects are handled we foreground them—this has far-reaching effects. Reality multiplies.

If practices are foregrounded there is no longer a single passive object in the middle, waiting to be seen from the point of view of seemingly endless series of perspectives. Instead, objects come into being—and disappear—with the practices in which they are manipulated. And since the object of manipulation tends to differ from one practice to another, reality multiplies. The body, the patient, the disease, the doctor, the technician, the technology: all of these are more than one. More than singular. This begs the question of how they are related. For even if objects differ from one practice to another, there are relations between these practices. Thus, far from necessarily falling into fragments, multiple objects tend to hang together somehow. Attending to the multiplicity of reality opens up the possibility of studying this remarkable achievement.

Philosophy used to approach knowledge in an *epistemological* way. It was interested in the preconditions for acquiring true knowledge. However, in the philosophical mode I engage in here, knowledge is not understood as a matter of reference, but as one of manipulation. The driving question no longer is "how to find the truth?" but "how are objects handled in practice?" With this shift, the philosophy of knowledge acquires an *ethnographic* interest in knowledge practices. A new series of questions emerges. The objects handled in practice are not the same from one site to another: so how does the coordination between such

ology." But what about all the exceptions? What about Marxist sociologists who, in the same decade, claimed that there was a *class division* running right through medicine (see, e.g., Chauvenet 1978). Or, for that matter, feminists, who were active in drawing distinctions between those parts of medicine that they saw as good for women and others, against which they pressed charges (see, e.g., Dreifus 1978)? Not to forget the combinations between the two (e.g., Doyal and Pennel 1979).

It would be possible to shuffle them aside, claiming that those texts have been marginal. In general, I could say, a few exceptions aside, for quite a while medical sociology took the medical profession to be a unity. Or I could point to these exceptions as the initial steps at the beginning of a new era. This would require me to say that up to the seventies medical sociology took the medical profession to be a unity, a position that slowly began to change. But this would still leave me with some problems. What if a more attentive reading of Freidson's book shows that its primary concern is *not* the profession's unity, but its *closed character*? When one reads him on his own terms, Freidson seems primarily worried about the lack of outside audit or control on medical mistakes and failures. If I still wanted to quote him as someone taking the medical profession to be a unity, I would then have to show that the profession's unity and its closure are closely linked, or indeed depend on one another. If that argument were hard to make, then I would have to find some other book

objects proceed? And how do different objects that go under a single name avoid clashes and explosive confrontations? And might it be that even if there are tensions between them, various versions of an object sometimes depend on one another? Such are the questions that will be addressed in this book. I cautiously try to sketch a way into the complex relations between objects that are *done*.

This book tells that no object, no body, no disease, is singular. If it is not removed from the practices that sustain it, reality is multiple. This may be read as a description that beautifully fits the facts. But attending to the multiplicity of reality is also an *act*. It is something that may be done—or left undone. It is an intervention. It intervenes in the various available styles for describing practices. Epistemological normativity is prescriptive: it tells how to know properly. The normativity of ethnographic descriptions is of a different kind. It suggests what must be taken into account when it comes to appreciating practices. If reality doesn't precede practices but is a part of them, it cannot itself be the standard by which practices are assessed. But "mere pragmatism" is no longer a good enough legitimization either, because each event, however pragmatically inspired, turns some "body" (some disease, some patient) into a lived reality— and thereby evacuates the reality of another.

This is the plot of my philosophical tale: that *ontology* is not given in the order of things, but that, instead, *ontologies* are brought into being, sustained, or allowed to wither away in common, day-to-day, sociomaterial practices. Medi-

to support my generalization. But which one? The problem is that many titles in medical sociology would do, in one way or another. There is a large corpus of texts in which the medical profession's unity is mentioned. But almost all of them, like Freidson's study, have other concerns at their core.

This is the point: generalizations about "the literature" always draw together disparate writings that have different souls, different concerns of their own. Stressing, in general, that *the* literature is attuned to medicine's unity may function to mark the originality of *this* study, a study that emphasizes disunity. But various dangers follow. One is that a false novelty is claimed: the ancestors are erased from memory instead of honored. A second is that, in the

case of this specific book, such generalities would create a tension between the ways in which "the field" and "the literature" are treated. If I take so much trouble to point out the multiplicity of medicine while I refer to sociology, anthropology, history, or philosophy in general terms, this might suggest that *they* possess the unity that medicine does not. But they don't. Just as it is possible to write about the multiplicity of the objects of medicine, this could be done about other disciplines. I won't attempt to do so here. But I will try to do justice to the variety of concerns, materialities, styles, and object framings in the various knowledges mobilized here by seeking not to suppress or hide these while relating to the literature.

cal practices among them. Investigating and questioning ontologies are therefore not old-fashioned philosophical pastimes, to be relegated to those who write nineteenth-century history. Ontologies are, instead, highly topical matters. They inform and are informed by our bodies, the organization of our health care systems, the rhythms and pains of our diseases, and the shape of our technologies. All of these, all at once, all intertwined, all in tension. If reality is multiple, it is also political. The question this study provokes is how the body multiple and its diseases might be done *well*. This question will not be answered here. Instead, I'll map out the space in which it may be posed.

The Perspectives of People

This is a philosophical book of a specific, that is, empirical, kind. It draws on social scientific and, more notably, ethnographic methods of investigation. But it does not just import these, it also mingles with them. For if I use ethnographic methods here, it is to study *disease*. That physicalities may be studied ethnographically is a quite recent invention. For a long time, "disease" was the unmarked category of anthropology and sociology of medicine. As the state of a physical body it was an object of biomedicine. Doctors told the truth about disease, or at least they were the only ones able to correct each other in so far as they didn't. Social scientists were careful not to get mixed up in this body-talk. Instead, they had something to tell in *addition* to existing medical knowledge. They pointed out that the reality of living with a disease isn't exhausted by listing physicalities. There is more to it. Apart from being a physical reality, having

Dates and Outdating

The work of Talcott Parsons is outdated. It is functionalist in character. *The Social System* is the title of his famous book of 1951 (Parsons 1951). It takes every social phenomenon to either be a threat to the system's stability or to have a stabilizing function. In chapter ten, "Social Structure and Dynamic Process: the Case of Modern Medical Practice," the social phenomenon analyzed in this way is *the sick role*. In modern society, Parsons argues, being sick is ritualized in a specific role. The sick don't need to work in the usual way but are, instead, taken care of. It is accepted that they are the victims of their sickness. This is good for society because if people stop working and take rest when they are sick this lowers the risk that they will die prematurely. In this way the chance that society has invested in someone's upbringing and education with too little return is reduced. However, since escaping from the usual obligation to work means that "being sick" may also be attractive, there is a potential threat. If everybody were to stop working by calling themselves sick, the system would collapse. This is why, in addition to withdrawal from work and being excused for such passivity, "the sick role" has two more elements. The patient has to go to bed and generally do whatever needs to be

a disease has a *meaning* for the patient in question. A meaning that is open to investigation. Listen to the story about Mr. Trevers (an invented name; all names used in field stories are invented):

Mr. Trevers sits in a chair in the surgical ward. Sure, he's quite willing to answer a few questions. Jeannette, the interviewer, sits down next to him. She casually asks if putting on the tape recorder is a problem. No, it isn't. They talk about the wound on Mr. Trevers's foot. It was the reason for the operation on his leg arteries a few days before. "My problem was not that it hurt," Mr. Trevers says, "but that this wound didn't go away. It was quite frightening. This gaping hole. I didn't go to the doctor at first, when that beam fell on my foot. I didn't care about the pain. But when it never went away, my wound, but only became bigger, then I got scared. And I went to see my general practitioner. She sent me in to the hospital. And now I've got two diseases. I've got atherosclerosis, they tell me, and diabetes. I've also got diabetes."

Mr. Trevers became frightened when his wound didn't heal. To the vascular surgeon who has operated on him, this fear is hardly relevant. It is relevant that Mr. Trevers finally decided to go and see a doctor. But once he did, well, fear, this is "one of the things people feel," as is an aversion of wounds that stay "gaping holes." If there is time, Mr. Trevers may be allowed to talk about his feelings. But they need not be written down in his surgical files. As "a good doctor" the surgeon may explain some facts in an attempt to reassure his patient. But "fear" is not a part of Mr. Trevers's vascular disease, nor of his diabetes.

done in order to recover. And the patient has to call on and follow the orders of a doctor who must officially sanction his or her sick role with a diagnosis.

This is functionalism: the sick role is described as a role that consists of four elements, which are all explained in terms of the function they have for the social system. Two of the role elements have a good function but risk undermining the social system, a danger the other two must counter. Overall, there is a balance between undermining and protecting elements, and the system maintains itself: it remains stable. In the fifties functionalism was strong, but it has been thoroughly undermined by later sociologists. By Marxists, who pointed out that functionalism forgets about antagonism, struggle, and change. By quantitative studies in which variables were isolated from each other and then correlated into causal chains, not functional schemes. By microsociologists, who pointed out that the many activities people engage in do not necessarily add up to form a stable whole, but point in various directions. And so on.

A lot more has changed in medical sociology since Parsons's time. Later medical sociologists still saw doctors as people who have the power to call a patient either "sick" or "healthy." Freidson, Zola, Szasz —they all insisted on this. But in their

As a complement to this, social scientists have made it their trade to listen for feelings when they interview patients. And they have persistently and severely criticized doctors for neglecting psychosocial matters, for being ever so concerned about keeping wounds clean while they hardly ever ask their patients what being wounded means to them. In addition to attending to blood sugar levels, bad arteries, wounds, and other physicalities, or so social scientists have been arguing in all kinds of ways, physicians should attend to what patients experience. This is how they have come to phrase it: in addition to *disease*, the object of biomedicine, something else is of importance too, a patient's *illness*. Illness here stands for a patient's interpretation of his or her disease, the feelings that accompany it, the life events it turns into.

In the social sciences, "disease" and "illness" were separated out as two interlinked but separate phenomena. Social scientists put "illness" on the research agenda. Shelves of books and volumes of journals were dedicated to it. Interviews were amassed, the attribution of meaning was analyzed, and ways of therapeutically attending to it were designed. All along social scientists left the study of disease "itself" to their colleagues, the physicians, until they started to worry about the power a strong alliance with physical reality grants to doctors. Then, social scientists gradually began to stress that reality isn't responsible all by itself for what doctors say about it. "Disease" may be inside the body, but what is said

work, the label "sick" was no longer presented as a potential favor a doctor may grant a patient, a good excuse to stop working temporarily. Instead, it was taken to be a negative judgment. A form of disapproval. In the 1960s the label "sick" came to be seen as a secularized form of the label "sinful." If doctors stick this onto people, they are being negatively labeled. So it wasn't only that functionalism became outdated. The label "sick" also changed from a kind of excuse or justification into a form of condemnation. And there is more. The kinds of examples used have also shifted. In Parsons's work, the implicit example is the infectious disease from which one either dies or fully recovers. The labeling theories that followed were concerned with forms of deviance like homosexuality and unmarried mother-

hood, which were called "sinful" in the forties and "sick" in the sixties. And after that came other examples: diseases caused by work or stress or social isolation. Chronic illnesses. AIDS. Reproductive technologies. So-called genetic diseases. One topic made way for another—though always only partially.

There are various layers of history to explore, and they all cover up Parsons and render him outdated. So why would one want to relate to his writings at all? The answer is that Parsons invented medical sociology. The crystallization of both of the objects of this discipline can be traced in his work. There they are: *illness* and *health care*. Let's look at chapter ten of *The Social System* again for their early articulation. Parsons links up with the broad definition of health that was popular in the

about it isn't. Bodies only speak if and when they are made heavy with meaning. In Mr. Trevers's case, a wound that doesn't heal is said to be a sign that points toward diabetes and atherosclerosis of the leg arteries. But this isn't necessarily so: this is a *meaning* that has been attributed. Such attributions have a history, and they are culturally specific. This opens them up for historical and social scientific investigations.

In this semantic approach, social scientists no longer primarily take doctors to be colleagues, colleagues who may be criticized for not listening to their patients carefully enough, but into whose object domain—physical reality—the social scientist wouldn't dare to venture. Instead of colleagues, doctors become the social scientists' objects. What doctors say when they talk about diseases is investigated with the theoretical tools that were crafted for studying the words of patients. Like patients, or so it is said, doctors have a *perspective*. They attribute meaning to what happens to bodies and lives. Even though doctors interpret the bodies and lives of others, whereas patients talk primarily about their own.

Perspectivalism turns doctors and patients into equals, for both interpret the world they live in. But to say this is also to reinforce their division, because the interpretations doctors and patients give must differ, linked as they are to the specific history, interests, roles, and horizons of each group. In perspectivalism, the words "disease" and "illness" are no longer used to contrast physical facts

medical world of his day. Health pertains to "a total physical, mental and social well being." (The definition was drafted at the first meeting of the World Health Organization in 1948 in a postwar atmosphere where hopes were high, but that's another story.) Illness, Parsons writes, "is a state of disturbance in the 'normal' functioning of the total human individual, including both the state of the organism as a biological system and of his personal and social adjustments. *It is thus partly biological and partly socially defined*" (431). A whole tradition of sociological thought can be traced back to this single sentence. Thus, Parsons's work is crucial. I must relate to it. Doing so may help me to escape from the coordinates in which it has been set.

How does Parsons frame these coordinates, how does he lay the grounds on which subsequent medical sociology was to move? The part of illness that is defined biologically falls under the technical competence of physicians. "Caring for the sick is not an incidental activity of other roles—though for example mothers do a good deal of it—but has become functionally specialized as a full-time job" (434). Parsons is not going to quarrel with physicians over their "functionally specialized" job. Their *technical competence*, that is, their knowledge of diseases and their skills in dealing with them, is granted to them. That's their domain. The sociologist, however, takes it to be his *own* technical competence to talk about jobs, functions, and specialisms. Refraining from talk about the contents of the physicians' competence allows Parsons to analyze the *social role of the physician*.

with personal meaning. Instead, they differentiate between the perspectives of doctors on the one hand and those of patients on the other.

Social scientist commenting on a talk I gave: "So you are going to tell us about their differ-ent perspectives on atherosclerosis, are you? Who are you going to include: only surgeons and radiologists? Or also internists, cardiologists, and general practitioners? You should think about epidemiologists, too; they tend to have an entirely different perspective. And what about nurses, you seem to be skipping the nurses, aren't you? And if you ask me, I think you'd better attend a little more to the patient's perspective, for that, after all, always receives far too little attention. And patients are what medicine should be about."

Like patients, professionals may be supposed to have perspectives of their own. However, these are *not* what I'll be telling about. There are, or so I want to argue, some problems with this line of work. It may seem that studying "per-spectives" is a way of finally attending to "disease itself"—but it isn't. For by entering the realm of meaning, the body's physical reality is still left out; it is yet again an unmarked category. But the problem has grown: this time the body isn't only unmarked in the social sciences, but in the entire world they evoke. The power to mark physical reality, after all, is no longer granted to medical doc-tors, it is granted to nobody. In a world of meaning, nobody is in touch with the reality of diseases, everybody "merely" interprets them. There are different in-terpretations around, and "the disease"—forever unknown—is nowhere to be

According to Parsons, the social role of the physician is "universalistic, func-tionally specific, affectively neutral and col-lectively oriented" (434). I will not com-ment on this specification of the "role of the physician" right now. I want to stress, instead, that Parsons turns this role into an object of sociological analy-sis. Something that sociologists may talk about, that is part of *their* technical com-petence. Later sociologists disagreed with Parsons—be it about the specificities of the physician's role, or about the question whether "role" is an appropriate term at all. In their disagreement, however, they all occupied this newly created space. A space from where they could speak with-out necessarily bothering about the self-understanding of physicians (and, later,

others engaged in health care practices). A space from which they could talk about the social specificities of health care—so long as they paid the price of refraining from talking about the technicalities of healing.

But then again. What is it to heal? Along-side the biomedical aspects that fall under the technical competence of the physician, there is the socially defined part of illness. Socially, being sick equals taking on the sick role. Parsons gives an analysis of this role that consists of four elements—but we've been there already. I've laid out the sick role for you while explaining Parsons's functionalism. There is, however, a good reason to come back to it. For if Parsons talks about "the sick role" *as a sociolo-gist*, then sociologists apparently have the

found. The disease *recedes* behind the interpretations. In a world of meaning alone, words are related to the places from where they are spoken. Whatever it is they are spoken about fades away.

Or maybe not. And that is the second problem of perspectival tales. In talk about meaning and interpretation the physical body stays *untouched*. All interpretations, whatever their number, are interpretations *of*. Of what? Of some matter that is projected somewhere. Of some nature that allows culture to attribute all these shapes to it. This is built into the very metaphor of "perspectives" itself. This multiplies the observers—but leaves the object observed alone. All alone. Untouched. It is only looked at. As if it were in the middle of a circle. A crowd of silent faces assembles around it. They seem to get to know the object by their eyes only. Maybe they have ears that listen. But no one ever touches the object. In a strange way that doesn't make it recede and fade away, but makes it very solid. Intangibly strong.

Is it possible to tackle these problems? Here's the task I've set myself. That is why I will *not* tell about the perspectives of medical doctors, nurses, technicians, patients, or whoever else is concerned. Instead I'll try to find a way out of perspectivalism and into disease "itself." How might this be done? By taking a third step. The first step of the social sciences in the field of medicine was to delineate *illness* as an important object to be added to a disease's physicalities. The second step was to stress that whatever doctors say about "disease" is *talk*, that it is part of a realm of meaning, something relative to the specific perspective of the person talking. And here is the third step. It consists of foregrounding

technical competence to talk about the socially defined part of illness. This means that sociology isn't simply outside health care, taking this as its object. Sociology also has the ability to acquire knowledge that may be useful to physicians. "If it ever becomes possible to remove the hyphen from the term 'psycho-somatic' and subsume all of 'medical science' under a single conceptual scheme, it can be regarded as certain that this will not be the conceptual scheme of the biological sciences of the late nineteenth and early twentieth centuries" (431). When "the hyphen" is removed, something will be added to biological science: sociological and psychological insights. Parsons's sociological insights, for instance. They tell, after all, that there are attractive sides to the sick role, and thus that doctors have to be careful, since a *motivational component* may be involved in getting sick.

There are still introductions to medical sociology that do not present "the sick role" as an outdated theoretical concept, but as a part of social reality. They relate to Parsons's work as if it belonged to the present. In up-to-date books and articles, however, other words are used. We'll come across an array of them. What they have in common, however, is their starting point. The idea that there is *more* to say about

practicalities, materialities, *events*. If we take this step, "disease" becomes a part of what is done in practice.

Reality in Practice

Let's move to hospital Z in order to learn about practice. In the ward where vascular patients are nursed, Jeannette sits on a chair next to the table by the window. She talks with Mr. Gerritsen. They have a long conversation, for the interviewer doesn't stick to the topics scribbled down in her notebook, topics that are not to be forgotten. Instead, she allows the patients to talk about whatever they want—or almost. The conversation may go in many directions, indeed, to the point where some patients get confused about the role they're playing and the game they've entered, and say things like "but you don't want to know all this, do you?" or "now what did you say this interview was for?" Mr. Gerritsen, however, shows no such hesitation. He seems pleased to talk. He's got so much to tell.

Mr. Gerritsen is sixty-two. Now that he's got an early retirement from his job and his daughters live on their own, he expected to have a quiet time. He thought he deserved it. He has taken care of his wife all the time she was ill, more than seven years, from the onset of her cancer till the day she died. His girls were still young then; they were 13 and 11 when his wife died. He didn't remarry, but brought them up by himself. And then, a few years ago, his legs

sick people than is told in biomedicine. The idea that there is a domain of "personal and social adjustments" that does not derive from physical facts, but has a specificity of its own. In that sense, medical sociology still largely moves within the coordinates set out by Parsons.

My point in relating this is not to say that Parsons was there first. In medical and social science journals of his time there are many texts that articulate comparable propositions for attending to *the social* in addition to attending to the body. (Parsons himself footnotes a physician/physiologist: L. J. Henderson, "Physician and Patient as a Social System" [Henderson 1935].) Authorship and origin stories don't concern me here. But Parsons is clear, sharp, articulate. Relating to

his work allows someone fifty years later to discover how the social sciences established their rights to speak about health care and sickness in the 1950s—and at the same time how they set limits on this right. They turned the domain of *the social* into what they were competent to speak about. In this way the social sciences delineated an object of their own *and* granted biomedicine the exclusive right to talk about the body and its diseases. So if I relate to Parsons's outdated work here, it is because, at the very moment of medical sociology's invention, he articulated so clearly the outlines of the place subsequently occupied by medical sociology for a long time. The place from which this book (like various others that appear these days) tries to escape.

started to bother him. As they do now. And it isn't that he's frightened or depressed by that, at least that's not the story he tells us. What Mr. Gerritsen talks about is the practicalities of living with legs that refuse to carry him. That hurt. This raises all kinds of difficulties. With the housework. The shopping. Social life. He explains them in some detail. Here's one example out of a long list of difficulties.

"My daughter, my eldest, she moved to another place. She went to live somewhere upstairs. She's still a young woman, so she's not going to get a flat on the ground floor. I haven't visited her yet. I can't climb all those stairs. 'Dad, when are you coming?' she asks. The child understands. She's not going to put pressure on me or something. But you want yourself to go there, don't you? I do. The problem is: it's four flights of stairs. Four. Yet there must be a way to do it. So I said to them, I said, 'what about putting a rope around my neck and then from up the stairs, from above, you pull and pull me . . . or . . . well, no' [laughs]."

This story reveals something about Mr. Gerritsen, all right. About his feelings, his sense making, and his self-irony. But in telling about the way he lives with his painful legs, Mr. Gerritsen also presents us with insights into the *events* that happen to someone with an impaired body. He tells about adapting his habits to his inability to walk without pain. And he tells about the limits of such adaptations. It may be possible to buy a small cart on wheels and so still be able

Aligning and Contrasting

In 1981, Allan Young published "When Rational Men Fall Sick: An Inquiry Into Some Assumptions Made By Medical Anthropologists." It is a critical article. Too many medical anthropologists, Young wrote at that time, listen to people talking about their sickness as if these people were "rational men." As if they presented theoretical knowledge and aimed to think along rational lines of argument. In contrast to these assumptions, Young argues that sick people's talk is usually of a different kind. When they talk, sick people do not necessarily refer to an internal state, but may instead attempt to reach some goal. They tend to not be in an operational mode, but in a preoperational one. This means that their own experiences are more important to them than any theory that might explain these experiences, that the words they use

may have an array of diverging meanings instead of a single one, and that their reasoning is not linear.

How to relate to this article? I might *align* with Young. He is right: anthropologists should not expect sick people to have "explanatory models." Patients should not be investigated as if they were the wild variety of the same species to which (domesticated) doctors also belong. The talk of sick people is a lot more complicated than the "rational man" scheme can hope to grasp. However, I could also be critical and *distance* myself from Young. Although he argues that sick people are not "rational men," he takes it for granted that doctors, by contrast, are. He takes them to be "operational" thinkers. However, it may well be that doctors are as nonlinear, complex, and self-contradictory as anyone else. (Here I might quote later studies

to do one's own shopping. But there are things, like visiting one's daughter, that become impossible, if she lives on the fourth floor, and one doesn't want to be hung on a rope.

Living with legs that hurt when walking does not only invite a person to make sense and give meaning to his or her new situation, but it is also a practical matter. A social scientist who wanted to know about the practicalities of living with bad leg arteries could follow Mr. Gerritsen while he does what he can and bumps up against what he cannot. Jeannette and I didn't undertake such an ethnography. But it is still possible for us to get to know some of the things we would have seen if we had followed him in his daily routine. We can listen to Mr. Gerritsen as if he were *his own ethnographer*. Not an ethnographer of feelings, meanings, or perspectives. But someone who tells how living with an impaired body is *done* in practice.

The stories people tell do not just present grids of meaning. They also convey a lot about legs, shopping trolleys, or staircases. What people say in an interview doesn't only reveal their perspective, but also tells about events they have lived through. If you agree to go along with this possibility for a while, and listen to patient interviews in a realist mode, the question becomes "what are the events people report on?" Here are three fragments, from three different patient interviews.

that investigate physicians' *thinking* and use these to support my argument; for example, Robert Hahn, "A World of Internal Medicine: Portrait of an Internist" [1985].)

And then there is a third way of relating to Young's 1981 article. This is to take out a half hidden remark, inflate it, and run away with it. At one point Young mentions that while people's talk isn't only cognitive, neither is cognition confined to talk. There is such a thing as *embedded knowledge*. This knowledge cannot be deduced from people's talk. It is incorporated in nonverbal schemes, in clinical procedures, in apparatuses. The precise formulations suggest that Young takes the existence of *embedded knowledge* to indicate a lack: a lack in the ability or willingness of those involved to be more articulate. He says,

for instance, that clinicians have a "tendency to continue investing in this knowledge rather than to self-consciously examine it," because "from the clinician's point of view, embedded knowledge is a form of professional investment (absorbing money and professional time) and is integral to his reputation and capacity for continued work" (324). Doctors do not take the time and effort to make things explicit, especially when this might make them more vulnerable. Rather, they spend their time working along.

Perhaps this is right. But it may also be that doctors do not "prefer" to invest in work rather than being explicit. It may be that this is not a matter of discretion. That, in other words, working in, by, and through embedded knowledge is simply the way

"I've got this nice neighbor, she's a young woman. So she takes me along. And then we do the shopping together. On Saturdays. With her car."

"But the scaffolding, and stairs, if you had to go some ten or twelve meters up, I couldn't do it any more. And walking, I went by bicycle, but in the end that became too much as well, if there's wind and you have to put in some effort. So my boss, well, he said, let's do it the sensible way, and he gave me things to do in the office. Until I could get an early retirement. So now I'm on early retirement. [Sigh] Yes."

"Then you're there at home. Alone. The whole day. And then I shuffle over the floor. I get to the sink all right. I can do it."

In the interviews we held with them, patients gave us detailed descriptions of the way they reorganize their households, their work, their family life. They told about how to get into the car or when to take a taxi. About steps and stairs, bicycles, and dogs on a lead. About the trials and tribulations of dealing with an impaired body in daily life.

But daily life isn't just located in houses, on streets, and in shops. Most patients also have lots to say about what happens to them in the hospital. It begins with going there. Can your son, who's so busy, take a day off to drive you? Where to find this room F021 where you've got to go? And are you allowed to

things are usually done. Everywhere—or almost. This would imply that cognitive operations are not central to what happens in hospitals. And, thus, that unraveling medical knowledge requires an investigation into clinical procedures and apparatuses rather than into the minds and cognitive operations of physicians. And, if one follows the argument through, that it might be better when interviewing people (whether they be doctors or patients) to ask them about what they do and about the events that happen to them, rather than about their thinking.

To take up the theme of "embedded knowledge" and run away with it is a more subtle way of relating to the literature than either agreement or criticism. Instead it sets up a *partial connection* between Young's text and this one. The connec-

tion is formed by the notion "embedded knowledge" minus the negative connotations Young attributes to it. So here is a link between my work and this specimen of the anthropological tradition. I investigate what Young calls "embedded knowledge" even if I use a different terminology.

But should I relate to this one article at all? It is a fine article, but Young has written so much more that is equally good. And so have others. There is a problem in making relations to the literature explicit: it takes so much space to outline a single, simple link such as this. And there are so many of them, so many more. They are *embedded*, indeed, in the questions asked, the topics raised, the words used throughout this study. Is it possible to make all the partial connections between a text and its relevant others explicit? I don't think so. So many

have anything to drink at all on the day of that examination with the difficult name? It all matters. It may all be told—or skipped.

Mrs. Gomans had an operation four days ago. Her left leg was opened up. One of her leg arteries was stripped: the atherosclerotic plaque taken away. Jeannette asks her to say more about it. "[Yawns.] Oh, well, the operation itself, you don't remember a lot about that. I had general anesthesia. So you come round again, and you have a suture, a scar in your leg, or two scars. And if it's all right your blood flow is good again. And then it's all done, and after ten days you can go home. And hop, you're a whole new person. That's the normal procedure."

The point is not that patients are necessarily the best possible ethnographers of the events that make up their own lives. They may be bad ethnographers. Mrs. Gomans, for instance, only tells about her scars when asked about her operation. She doesn't go into detail, but talks about a "normal procedure" in terms heavily colored by her hope of becoming a "new person." But then: Mrs. Gomans had general anesthesia. And general anesthesia is a pretty strong method of turning someone into a bad ethnographer.

Other patients have seen more. They've tried hard.

resonating terms, so many diffracting topics, so many shared words with slightly different meanings from one text to another. How much time and energy should we invest in articulating relations—might it not sometimes be simply better to get on with it, and show some "capacity for continued work"? The question may be asked. But then, if articulating embedded knowledge is hardly a matter of discretion for doctors, one may wonder to what extent it is for theorists.

Disciplines

This book is about the multiplicity of the body and its diseases. Does that topic locate it within a specific *discipline*? Perhaps, but if so, then which: medical sociology? medical anthropology? or medical philosophy? I have a hard time thinking about where to locate myself, or better, where to locate this book. And the thinking

is the easiest bit. How to *do* location work? The classic way is by relating to the literature that constitutes a disciplinary field. Thus: by relating to Freidson or Parsons I turn this text into medical sociology. Relating to Young is a way of making a connection with medical anthropology. In due course I'll have to put in some philosopher or other—maybe Canguilhem is a good idea—in order to pass as a medical philosopher.

But there are so many other fields, places, literatures to relate to. Crucial among these are the dispersed studies about the framing of boundaries between biology and the social sciences. After the Second World War it became important to draw these boundaries and to draw them sharply. Talk about biological differences between humans had become suspect. In an attempt to make a sharp contrast with the murderous eugenic practices of the

Mr. Jonas also had an operation a few days ago. He had a local anesthetic, so he was conscious. And he was fascinated. "I could see quite a bit of the last operation. I found it interesting. First they hang some cloth right in front of your face. I said [to the anesthesiologist]: man, take that thing away. I said: I can't see anything like that. He said, no, he said, I can't do that, most patients can't take it, they get sick, he said, and we can't have that, during an operation. So I said, well, if by accident I pull that cloth away then you shouldn't be angry at me, I said. For I do want to see it for myself. And then I was allowed to see it. I saw three-quarters of the operation. Yes, I thought it was quite interesting. After all, it's your own body [laughs]."

Mr. Jonas was eager to observe the interventions to which his body was subjected. When asked what he saw, he tells that there were lots of people around, all in green and with caps covering their mouths and noses. They were handing each other instruments. And at one point the surgeon said something about amputation. Mr. Jonas realized almost immediately, he says, that this was in order to teach the students present about the risks of people with bad arteries who are not treated. It was only a fraction of a second that he feared it was his own leg that would be amputated.

Nazis, the biological equality of humankind was stressed. Talking about differences between humans became a privilege of the social sciences, even if this privilege had to be reconquered over and over again (see, e.g., Rose 1982). In various forms and variants the social sciences delineated their own objects *alongside* those of biology. One of the arguments for doing so was that this helped to ward off racism.

In 1982, Martin Barker published *The New Racism*. It presented an analysis of the discourse of the British New Right of that time. It showed that in the discourse of this movement the dominant talk was no longer about the inferiority of so-called other races. Instead of biological inferiority, *cultural difference* was said to mark the contrast between "us" and "them." "Their" bodies don't stink, but "their" food does. And "they" do not need to be eliminated: all "they" need to do is go home—

wherever that is. Barker's book isn't in the field of medical anything. But even so it is a good text to relate to because Barker makes it very clear that a division between biology and sociology that was once helpful may lose its earlier power. In the case he analyzes, it no longer does the political work it was called in to do, which is to counter racism. Racism may be grounded in culture as well as in nature.

In *After Nature: English Kinship in the Late Twentieth Century*, Marilyn Strathern takes distance from the nature/culture division in another way (1992a). She shows that anthropology has undertaken studies of *kinship systems* all over the world, as if kinship systems were social constructions building on natural facts. Kinship studies departed from the alleged natural fact that children are the offspring of a father and a mother, inheriting equal amounts from either parent. Strathern, however, makes

So Mr. Jonas saw a lot. But when Jeannette asked him what happened to his legs, he only points to the places where his skin was cut. There's where it happened. He adds that he remembers images, and vividly too, but he can't tell anything much about them. How should he? The terms with which one might articulate the specificities of an operation do not belong to his vocabulary.

Mrs. Ramsey has had her first operation. Jeannette asks Mrs. Ramsey whether she thinks she might need another one in the future. There are, after all, many patients around who've already been here four, five times before. The question, however, provokes a frightened reaction. "Oh, please, not again. No, I hope not. I hope never to have another operation. For they laid me down on this sloping table. They had to work on the leg's side. The table was completely tilted, sloping. So the way you are hanging there, all the time, in that thing. You're stiff from that, too, of course, in the end. And your muscles ache when you wake up. They do."

Mrs. Ramsey doesn't yawn and she isn't optimistic, like Mrs. Gomans. Nor does she express the remarkable curiosity of Mr. Jonas. Instead, her tone is anxious. She disliked undergoing an operation and the physical discomfort involved. But her aversion does not stop her from making several interesting observations. That the operation table was tilted. That this was related to the surgeons' task of reaching the side of her leg. That she was stiff afterward.

it plain that these "facts of life" were cultural all along. They were English, to be precise, expressions of the English kinship system of the twentieth century. The very terms in which the so-called natural facts of procreation were expressed already incorporated images of social relations, like the word "heritage," which was about inheriting wealth long before it was about inheriting genes. In order to take distance from her own Englishness, Strathern calls in the theoretical help of various Melanesian peoples, in whose conceptual schemes there are no nature and no culture. Instead, they know sons who give birth to their fathers. Women who are men in a female form and vice versa. Containers that are, in their turn, contained.

But it is not only Melanesian theories that help Strathern to escape from tradi-

tional English schemes. The current technical reshaping of human reproduction works in this way, too. Now that a newborn baby can be the genetic child of one mother and the anatomical child of another, the old schemes start to crumble. The opposition between a singular natural parenthood and a pluralist range of cultural constructions that shape it later on no longer holds. What will come next? Western/English culture is, in a material way, by changing nature, undermining the grounds of its own nature/culture divide.

And there is a third parallel move that I would like to relate to. It is not about race or kinship, but about sex. Sex and gender. Here's a quote from Donna Haraway's essay "Gender for a Marxist Dictionary: The Sexual Politics of a Word": "In 1958, the Gender Identity Research Project was

It is possible to listen to people's stories as if they tell about events. Through such listening an illness takes shape that is both material and active. It is an illness that consists of lying on a sloping table. Of arguing with your anesthesiologist about the cloth in front of your eyes. It is an illness made up of scars on your legs that do not stop you from becoming a new person. This illness is something being done to you, the patient. And something that, as a patient, you do.

Who Does the Doing?

The patient stories quoted above do not expel physical reality. Instead, they talk about it, for it is everywhere. The physicality of bodies, vessels, blood. That of shopping, trolleys, and staircases. And that of anesthetic drugs, green clothing, knives, and tables. What is important about this reality is that it hurts, makes noises, smells. That it bleeps or falls on the floor. That it is touched. Patients may interpret bodies, but they also live them. And so do doctors. Doctors figure prominently in the patient stories. They administer general anesthesia. They are dressed up in green, use instruments, cut open legs, and close them again with thread and needle. They teach, are taught. They tilt tables and work on the inner sides of legs. They do lots of things to the patient's body.

Perspectivalism puts doctors and patients on a par, with a great divide between them, because they cast their views from different angles. The traffic

established at the University of California at Los Angeles (UCLA) Medical Center for the study of intersexuals and transsexuals. The psychoanalyst Robert Stoller's work discussed and generalized the findings of the UCLA project. Stoller introduced the term 'gender identity' to the International Psychoanalytic Congress at Stockholm in 1963. He formulated the concept of gender identity within the framework of the biology/culture distinction, such that sex was related to biology (hormones, genes, nervous system, morphology) and gender was related to culture (psychology, sociology)" (Haraway 1991, 133).

Another division articulated in the fifties: one that made it possible to talk about a person's "gender" irrespective of the biological "sex" of her/his body. And it wasn't only intersexuals, transsexuals, and their therapists who talked in this way. Feminists came to do so, too. It helped them to fight the biological determinism that tried to put women in subordinated positions using their female bodies as a legitimization. Haraway, however, wants us to get away from the sex/gender division because by going with this distinction feminists may have gained the right to talk about the social shaping of gender "irrespective of sex"—but the price they paid was that they left the category "sex" unanalyzed. Unquestioned. "Thus, formulations of an essential identity as a woman or man were left analytically untouched and politically dangerous" (134). If we want to

across the doctor-patient divide attracts much public attention. There are fascinating books written by doctors who describe the way their perspective changed when they fell ill and became a patient. Social scientists investigate whether and how patients incorporate medical schemes and terminology into their own thinking. Conversational analysis shows how the boundary is — or isn't — crossed when doctors and patients interact. And testimonies of patients who relate the inability of their doctors to understand them make for disheartening reading. But both the difficulties *and* the possibilities of crossing the gap point toward the existence of that gap. So there it is. A cleavage. A perspective from one point of view differs from that of the other.

In stories that tell about events-in-practice this is different. However shared or solitary perspectives may be, the practice of diagnosing and treating diseases inevitably requires cooperation.

The surgeon walks to the door and calls in the next patient. They shake hands. The doctor points at my presence and says that I'm there to learn something. He sits on a chair behind his desk. The patient, a woman in her eighties, takes a chair at the other side of the desk, clutching her handbag on her lap. The doctor looks in the file in front of him and takes a

analytically touch the way woman and man are framed in biology, we need to address the issue.

In the texts of Barker, Strathern, and Haraway I relate to here, *the words* "disease" or "illness" never appear. And yet these texts are worth relating to when I want to argue that we should no longer leave the study of "disease" to biomedicine. There are analogies. Differentiations between race and culture, biological parenthood and kinship systems, sex and gender, or disease and illness have a lot in common. Each of these distinctions was crafted in the 1950s in order to create a space for the social sciences alongside biology. *Alongside:* the metaphor may be taken seriously. The social domain was regionally separated from the biological domain. This both solved and created problems. Problems that were comparable.

Of what kind? Barker shows that making

space for "culture" next to "nature" is no longer a protection against racism. Racism can be framed in cultural as well as natural terms. Parallel to this, one might conjecture that it isn't only knowledge of "disease" that holds power over patients these days. Knowing "illness" also does; for example, the "quality of life" so important in framing present-day health care is defined in sociological terms. Strathern tells that the nature/culture divide is subordinate to its latter element: it is the invention of a quite specific culture. Where this very culture is changing its procreative possibilities, new schemes will (have to) emerge. Translating this analysis, it is possible to say that one of the dominant ways Western cultures live their "illnesses" is by taking them to be "diseases." Things doctors know about. But recent transformations in health care, like those that make patients into the guardians of their own

letter out. "So, Mrs. Tilstra, here your general practitioner writes you've got problems with your leg. Do you?" "Yes, yes, doctor. That's why I come here." "Tell me, then, what are those problems? When do you have them?" "Well, what can I say? It's when I try to do something doctor, move, walk, whatever. Like, I used to walk the dog for long stretches, but now I can't. I hardly can. It hurts too much." "Where does it hurt?" "Here, doctor, mostly down here, in my calf it does. In my left leg." "So it hurts in your left calf when you walk. Now how many meters, if you walk on flat ground, say, how many meters do you think you can walk before it starts hurting?" "What can I say? I think it must be, well, some, not a lot, some fifty meters I guess." "Good. Or not good. Well. And then, can you walk again, then, after some rest?" "Yeah, if I wait for a while, after that, yes. I can, yes."

In the consulting room something is *done*. It can be described as "pain in Mrs. Tilstra's left lower leg that begins on walking a short distance on flat ground and stops after rest." This phenomenon goes by the medical name *intermittent claudication*. Whatever the condition of her body before she entered the consulting room, in ethnographic terms Mrs. Tilstra did not yet have this disease before she visited a doctor. She didn't *enact* it. When all alone, Mrs. Tilstra felt pain when walking, but this pain was diffuse and not linked up to a specific walking

therapies, are in the process of undoing the former divisions.

Haraway, finally, warns that feminists who fight biological determinism in a dichotomizing way, one that puts the body's "sex" in a safe domain before starting to discuss a person's "gender," leave biology unanalyzed. " 'Biology' has tended to denote the body itself, rather than a social discourse open to intervention" (134). Thinking along with this, in an activist mode, one might say that leaving "disease" in the hands of physicians alone is a political weakness. For whatever one may say about the social shaping of the former sick role, whatever one may say about "illness," as long as "disease" is accepted as a natural category, and left unanalyzed, those who talk in its name will always have the last word. It would be better to mix with them, move among them, study them, engage with them in serious discussion.

By relating to it, I may try to *import* the work of Barker, Strathern, and Haraway into the fields of medical sociology, anthropology, or philosophy. Importing texts from other fields tends to be a good way to say "new" things. But where are these texts coming from? Not from a clear-cut discipline, but from an interdisciplinary, slightly undisciplined field. A flow of theory moving across boundaries. The boundaries of disciplines, of nature and culture, of theory and politics. Maybe relating to them is a good way to give *this* text a place in that nondisciplinary fluid space as well.

Visibility and Access
Relating to the literature may be a way of situating one's own text among others, which tends to be helpful to most readers. It may be a way to sketch the ancestry one is shaped by and the elders one seeks to depart from, and these may or may not

distance on flat ground. The trouble Mrs. Tilstra encountered when she tried to walk her dog did not yet have the shape that emerges when she answers her doctor's questions.

This does not imply that the doctor brings Mrs. Tilstra's disease into being. When a surgeon is all alone in his office he may explain to the visiting ethnographer what a clinical diagnosis entails, but without a patient he isn't able to *make* a diagnosis. In order for "intermittent claudication" to be practiced, two people are required. A doctor and a patient. The patient must worry or wonder about something and the doctor be willing and able to attend to it. The doctor must ask questions and the patient be willing and able to answer them. And in addition to these two people there are other elements that play a more or less important role. The desk, the chairs, the general practitioner, the letter: they all participate in the events that together "do" intermittent claudication. As does Mrs. Tilstra's dog, without whom she might not even have tried to walk more than the fifty meters after which her left leg starts to hurt.

Another scene.

Even if he has come all the way to the consulting room, Mr. Romer never gets to speak. His wife has come with him. She does the talking. "He's not doing well, doctor, he isn't. He

be the same. For all of these purposes it is best to relate to literature that has some authority. If I relate to Parsons, advanced readers are more likely to get at what I'm trying to do, for they know Parsons. And novice readers are well served as well, for they will have to get to know Parsons sooner or later if ever they want to be taken seriously. If I import Strathern my text becomes stronger, for whoever wants to argue against my playing around with the nature/culture divide now has to argue with her as well—and she's written a lot about it. But how do authors ever acquire authority? Answer: by being related to. It is a circle.

There is an interesting article by Robert Pool that I would like to relate to. Pool is a medical anthropologist who was sent out (and paid) to investigate why kwashiorkor, a disease of malnutrition, is com-

mon in the northwest region of Cameroon, even when there is enough food to go round. In order to start to answer this question, Pool tried to get insight into the way the people of the village where he lived talk about this disease. He wanted to explore their own perspective on it. Their interpretations, their illness story. But the stories that people in the village told Pool weren't about kwashiorkor. Or, if they were, they were also about a lot more. Talk shifted between *kwashiorkor, ngang,* and *bfaa*. There were no solid boundaries between these words. None of them was used in a way that stayed stable from one conversation to the next. And if it hadn't been for Pool asking questions, the stories wouldn't have been told (let alone in the way he recorded them) in the first place.

Pool makes it clear that if an anthropologist goes out to study "illness" as if it were

can't do a thing any more." "So, Mr. Romer," the surgeon says, trying to look the old man in the eye, "what's the problem? What do you come to see me for?" "It's his legs, doctor," Mr. Romer's wife answers. "He's had a heart attack, he's had two in fact. But now it's his legs. He can't get himself to walk any more. He has too much pain." Mr. Romer looks worn out. And despite the surgeon's stubborn attempt to address him, Mr. Romer doesn't speak. Maybe he can't. Maybe—the surgeon seems to reckon with that possibility—he has given up trying.

A doctor cannot diagnose intermittent claudication all alone. He needs others for it. But the scenario isn't rigid. Many of its elements are flexible. Instead of fifty meters, the walking distance may be a hundred meters. Instead of the calf, the thigh may hurt. And if the patient cannot speak, someone else may speak for him. But what is needed, indeed indispensable for clinical diagnosis, is that there be a patient-body. This must be present. And it must cooperate.

The surgeon looks from the file to the Romer couple and back down to the file, where he makes a few notes. His head up again, he says, "Now, if you please, Mr. Romer, I'd like to take a look. I want your legs, I want to see for myself what they look like. And feel your blood vessels. For you may have a problem with your blood vessels." After having said this in a loud voice, the surgeon turns his head to Mrs. Romer—thereby accepted as a spokes-person—and asks: "Do you think it's possible for him to take his trousers off and lie on the examination table?" It is possible but not for Mr. Romer all by himself. It isn't easy. The

the lay theory of a "disease" that doctors talk about in medical terms, he is trapped from the beginning. He is trapped in "disease" language. Why would laypeople, let alone in Cameroon, delineate entities in their own talk that nicely parallel the categories of Western medicine? To presume this is to presume that the disease categories of Western medicine are "natural." That they reflect a reality out there for everyone to stumble over before interpreting it in diverse ways. But people's categories, or so Pool argues, do not reflect a nature accessible, if in various ways, for all. Instead, they are part of a specific practice for dealing with life, suffering, and death.

The article mentioned is entitled "Gesprekken over ziekte in een Kameroenees dorp: Een kritische reflectie op medisch-

antropologisch onderzoek"(Pool 1989). Do you recognize the language? It's Dutch. There must be lots of interesting articles I cannot relate to because they are in languages I do not understand. In Danish, Italian, Urdu, Kishwahili. And while I happen to read Dutch, I don't know when to relate to Dutch literature very well when I'm writing in English. If I do, the risk is that it does not increase the transportability of my endeavor. It does not help you to situate my texts but, worse, is likely to be frustrating for you. My reference might make you eager to read Pool. Well, in this instance you can. He's also published a book in English, in which he develops the same arguments (Pool 1994). But there you are lucky.

limbs are heavy. Shoes and socks can only be undone when the feet are lifted. The zip refuses, the trouser fabric is stiff. Then there's the height of the table. But after a while the vascular surgeon holds Mr. Romer's two feet in his full hands to estimate and compare their temperature. He observes the skin. And with two fingers he feels the pulsations of the arteries in the groin, knee, and foot. "Can you flex your leg a bit for me, please, yeah, yes, that's it, there you go. Very good."

In their consulting rooms, vascular surgeons add a physical examination to an interview. The patient's answers to the diagnostic questions may make a typical story or a vague one. They may be enough to talk of intermittent claudication straight away, or not quite so. In either case, the enactment of intermittent claudication is extended and strengthened by adding the elements a physical examination may yield. Cold feet, or one cold foot. Weak pulsations. A thin, poorly oxygenated skin. To add such elements, the patient's legs and the doctor's hands cooperate. As do the examination table and the person who helps a patient worn out with age to undo his shoes and take off his socks and trousers.

Who does the doing? Events are made to happen by several people and lots of things. Words participate, too. Paperwork. Rooms, buildings. The insurance sys-

No Space/Time Available!

No text goes everywhere. In this book I do not go into the history of the diseases I describe. I even flatten out most of the changes observed over the few years of my fieldwork. The fact that there is difference over time is among the sacred truisms of current-day social theory. One of the attacks on functionalism was indeed that it could not account for change. That it draws spatial images of society, instead of making graphs about events over time. For a long time *process* has been such a buzzword that when doing (social) theory, one could hardly do without it. But in this book the matrices produced are primarily spatial. The different configurations to be mapped are *next to* one another, or *inside* or *above.* And I will also play with and shift Euclidean images of space and come to talk about such figurations as *mutual inclusion.* More about that later.

What is important to note now is that this book does not go into history. Should it relate to historical literature even so? There are, after all, lots of intriguing and relevant studies to find there. One of them is Barbara Duden's *The Women Beneath the Skin* (Duden 1991). It does theoretical work. Duden's history delves under the skin of human bodies. That makes it of immediate relevance to the social sciences as well as the philosophy of medicine. Duden makes her readers *feel* that the experience of one's own physicality from the inside does not precede culture. Not that just about anything could be produced or constructed, precisely not that: the flesh is stubborn. It is stubborn as long as it is alive, but even so it is a historical phenomenon. And its historicity is not a mere matter of interpretations changing, but of the very fleshiness of being alive itself.

tem. An endless list of heterogeneous elements that can either be highlighted or left in the background, depending on the character and purpose of the description. The descriptions given here are mine, not those of Mrs. Tilstra, Mr. Romer, or any other patient. And even if my descriptions are informed by what patients tell about events, I only rarely follow patients in this book. This ethnography (that is its force but also constitutes its limits) concentrates on medicine: it is made to unravel medical knowledge, medical technology, medical diagnosis, and medical interventions. It is informed by my own observations and by attending primarily to the words of another group of lay ethnographers: medical professionals.

The outpatient clinic. The patient who's next on the list never showed up. The one after that hasn't arrived yet. So we walk to the coffee machine and the vascular surgeon inserts his identity card and gets us two coffees. We stroll back to his office. Talk as if chatting. "You see, you've got to understand this," he says, wanting to be a good informant, "making a diagnosis is very different depending on whether or not they have a good general practitioner. Sometimes people come here and there's this letter and everything is in it: walking distance, pulsations, name it. A detailed history. Now of course you check that, you take every single step yourself again, but in such cases you may be fairly sure where you'll be going. But it also happens that there's just some illegible sentence scribbled down like 'please see this patient for me.' So then your work is different. Often, in such cases, there isn't a vascular problem at all. It may be something neurological. Or whatever. Nothing."

Duden presents an analysis of detailed reports of the complaints and wishes of female patients that a doctor in a small German town published in 1730. There are a lot of intermediaries between these women's physical experiences and the readers of Duden's book. The situatedness of the medical practice of the author, the specificities of his medical vocabulary, the writing habits of his time, Duden's own selection as a historian: readers get to know a lot about these. And yet from what we read, it seems inescapable that the bodies of these women were different from those that we inhabit now—however large the differences between us. We simply couldn't *do* such a body any more, nor describe it from the inside.

All these words that Duden lists meticulously! About hurt wandering through the body. Flows, white or red, that may flow out of womb or skin or eyes. The worry of the women that their blood is driven inside them, gets stuck there, sticks there, will not come out. Duden delves into the descriptions in her eighteenth-century material to come out with a lived body—but one that lives a life different from our own. Relating to Duden allows me to *import* this conclusion: even the lived experience of one's own body is mediated. It is not that just any form can be plastered into it. But neither is it the case that the *modern Western body* preceded medicine—subsequently to be objectified by it. They both have a history. These histories may well be intertwined.

I might never have noticed that the diagnostic work of a vascular surgeon differs considerably with the letters of general practitioners if it hadn't been for the conversation just quoted. Thus, in the ethnography engaged in here doctors become the social scientist's colleagues once again. They stop being "mere" objects of research whose interpretations may be listed and related to their historical and cultural context. But neither are they the colleagues they used to be, professionals who have knowledge of "disease," to which the social scientist may add knowledge about "illness." Instead, the territorial boundaries of professionalism are starting to leak. Doctors talking about their work may be listened to as if (like patients) they were their own ethnographers; ethnographers in their turn need not stop short as soon as they come across machines or blood, but can continue their observations. They may write about the body and its diseases.

In this unbounded territory, the disease/illness distinction is no longer helpful. When doctor and patient act together in the consultation room, they jointly give a shape to the reality of the patients' hurting legs. How to call what they thus shape? If I use the word *disease* here, this is not to locate my text on the disease side of the disease/illness distinction, but to breach it. To make it plain that I will attend to physicalities even if I am not a medical doctor. To underline that it can be done. That there are ways of ethnographically talking bodies. There are good reasons to try, if only this one: that the *humane* does not reside exclusively in psychosocial matters. However important feelings and interpretations may be, they are not alone in making up what life is all about. Day-to-day reality, the life we live, is also a fleshy affair. A matter of chairs and tables, food and air, machines and blood. Of bodies. That is a good reason not to leave these issues in the hands of medical professionals alone but to seek ways, *lay ways* so to speak, to freely talk about them.

Never Alone

In their outpatient clinic, vascular surgeons interact with patients. Here's what the doctors do: they ask questions (where does it hurt, how long can you walk, does it stop when you rest?) They look at the color and the texture of the skin of legs that hurt. They put their hands on places where the patients' leg arteries should be palpable and attempt to feel whether or not the arteries pulsate with each heartbeat. They scribble down notes in their files while their patients quickly or clumsily put on their clothes again. And then they propose the next step in the patients' itinerary. I've seen them doing this again and again, sitting on a stool with a white coat on, smiling, or looking serious. That's what the vascular surgeons of hospital Z showed me when I asked them about "atherosclerosis of the leg vessels": they took me to their outpatient clinic.

Then I wanted to know about pathology. The doors of the department of pathology say "No Entry." Being a researcher, I was kindly permitted to use them as an entrance even so. It was, however, not possible to see atherosclerotic leg vessels any random week there. The pathology resident who was to be my informant phoned me when he had something to offer. "I've got a leg," he said. A few days and preparatory steps later we finally saw what I had come looking for. Atherosclerosis.

In the small room he shared with two others, books and papers all around, the pathology resident had installed the double microscope for the occasion of my visit. "If I'm alone I use one with just a single pair of eyepieces," he said, "this one is used when a supervisor wants to check what we are doing." We sat down with the microscope on the table between us. Each of us looked into one of the eyepieces. He focused the image, asking me when what I could see was sharp. With an inbuilt pointer he taught me what to see. As if he were, today, the supervisor.

"You see, there's a vessel, this here, it's not quite a circle, but almost. It's pink, that's from the colorant. And that purple, here, that's the calcification, in the media. It's broken. They have done a bad job with the decalcification. Not done it long enough, so the knife had a problem cutting. Look, all this, this messiness here, that's an artifact from that." He shifted the pointer to the middle of the circle. "That's the lumen. There's blood cells inside it, you see. That only happens when a lumen is small. Otherwise it's washed out during the preparation. And here, around the lumen, this first layer of cells, that's the intima. It's thick. Oh, wow, isn't it thick! It goes all the way from here, to there. Look. Now there's your atherosclerosis. That's it. A thickening of the intima. That's really what it is."

And then he adds, after a little pause: "Under a microscope."

My endeavor hinges on this last addition. The pathology resident utters it as if he is saying nothing special. "Under a microscope." But it implies a lot. Without this addition, atherosclerosis is all alone. It is visible *through* a microscope. A thickened intima. There is something seductive about it. To bow one's head over a microscope and let one's eyes be directed by the pointer. If only be-

Studying Practice
In this book I reflexively attend to the genre of "relating to the literature." I am not all that comfortable with this genre, for there is the danger that it implicitly strengthens a number of assumptions against which the text is making explicit arguments. Besides, it is never possible to relate to the literature specifically enough. Out of sheer love for detail, I would prefer to not include any references at all, since they will inevitably be too crude. But that is not wise. "A paper that does not have references is like a child without an escort walking in the night in a big city it does not know: isolated, lost, anything may happen to it" (Latour 1987, 33). Presenting this quote is a way of relating to the literature that I will use only sparingly throughout this book: treating it as a source of authority. If Latour

says papers need references, then so they do, or would you want to disagree with him? And if papers need references, then so do books.

The reference to Latour that helps to introduce some of the background of the present study is to his *We Have Never Been Modern* (Latour 1993). In that book, Latour seeks ways out of the nature-culture divide —just like Barker, Strathern, Haraway, and many others whom I have not mentioned. Latour doesn't follow the way this divide was framed and institutionalized in the twentieth century, but, in a wider gesture, links it up with modernity. All modern thinkers, he claims, glorify their ability to distinguish between natural and social phenomena, disqualifying those who are "unable" to do so as premoderns. Meanwhile, however, or so Latour argues, in the

cause a vessel cross section makes for a beautiful image. With all its pink and purple and its strange forms that slowly come to be discernible if their nature is explained. There's something seductive about it: to use instruments as "mere" instruments that unveil the hidden reality of atherosclerosis.

But when "under a microscope" is added, the thickened intima no longer exists all by itself—but through the microscope. What is foregrounded through this addition is that the visibility of intimas *depends on* microscopes. And, for that matter, on a lot more. On the pointer. And on the two glass sheets that make the slide. Don't forget the decalcification that, even when it isn't done long enough, allows the technician to cut thin cross sections of a vessel. There's the work of that technician. The tweezers and the knives. The dyes that turn the various cellular structures pink and purple. They are all required if pathologists are to see the thick intima of a vessel wall.

It may be foregrounded or forgotten. When they talk bodies, doctors switch. Sometimes they add "under a microscope" or some equivalent of that. Sometimes they don't. My ethnographic strategy hinges on the art of never forgetting about microscopes. Of persistently attending to their relevance and always including them in stories about physicalities. It is with this strategy that disease is turned into something ethnographers may talk about. Because as long as the practicalities of *doing* disease are part of the story, it is a story about practices. A praxiography. The "disease" that ethnographers talk about is never alone. It does

---

practices of the so-called modern world the natural and the social are as intertwined as they are in so-called premodern *thinking*. This implies that there are clashes between the knowledge *articulated* in technoscience societies and the knowledges *embedded* in their practices. While the importance of a clear-cut distinction was loudly proclaimed, it wasn't converted into action. Therefore, *modernity* is a state we have never been in, for only our theories make modern divides. Our practices do not.

Latour addresses several versions of the natural/social divide. One of these is the distinction between *subject* and *object*. In the schematic models of modernity, or so Latour explains, the subject, which is social, actively knows, and the object, being

known, is natural. In order to overcome this divide we have to learn to realize that the world we live in is a mixture. Latour's way of achieving this is to claim that subjects and objects are two poles of a spectrum, which have many quasi subjects and quasi objects, mixtures, in-between them. The moral of the book is that instead of dialectically jumping between the ideas that reside in the minds of subjects and some objective reality *out there*, we would do better to admit that in our daily lives we are engaged in practices that are thick, fleshy, and warm as well as made out of metal, glass, and numbers—and that are persistently uncertain.

Relating to this statement allows me to explain to you better, I hope, what it

not stand by itself. It depends on everything and everyone that is active while it is being practiced. This disease is *being done*.

No, pathologists do not *make* the thick atherosclerotic vessel walls they look at, nor do they *construct* them. Those are clumsy words for what happens in the department of pathology of hospital Z. They suggest that material is assembled, put together, and turned into an object that subsequently goes out in the world all by itself. Instead of the "construction" metaphor of the workshop we might try to mobilize a theater metaphor for what happens in the hospital. When a disease is being done, we may say that it is *performed* in a specific way. The word "performance" has various appropriate connotations. There may (but need not be) a script available for doing a disease. If the script is not put to play, it is of no value for what happens in the theater. At different times and places scripts are staged in various ways. If there is no script, actors improvise. The stage props are as important as the people, because, after all, they set the stage.

But then again, the performance metaphor has some inappropriate connotations as well. It may be taken to suggest that there is a backstage, where the real reality is hiding. Or that something difficult is going on, that a successful accomplishment of a task is involved. It may be taken to suggest that what is done here and now has effects beyond the mere moment—performative effects. I don't want those associations to interfere with what I want to do here: to shift from an epistemological to a praxiographic inquiry into reality. So I need a word that doesn't suggest too much. A word with not too much of an academic history. The English language has a nice one in store: *enact*. It is possible to say

is I am trying to do in the present book. What I am attempting is similar. I investigate knowledge incorporated in daily events and activities rather than knowledge articulated in words and images and printed on paper. I privilege practices over principles and study them ethnographically. This turns *doing anthropology* into a *philosophical move*. A move away from the epistemological tradition in philosophy that tried to articulate the relation between knowing subjects and their objects of knowledge. The ethnographic study of practices does not search for knowledge in subjects who have it in their minds and may talk about it. Instead, it locates knowl-

edge primarily in activities, events, buildings, instruments, procedures, and so on. Objects, in their turn, are not taken here as entities waiting out there to be represented but neither are they the constructions shaped by the subject-knowers. Objects are—well, what are they? That is the question. That is the question this book tries to address.

So just as Latour in *We Have Never Been Modern* recommended, I want to escape the subject/object divide. But there is also a difference. I want to escape from this dichotomy *twice*. I will argue in what follows that it is not a *single* dichotomy; there are (at least) *two* subject/object divisions

that in practices objects are *enacted*. This suggests that activities take place—but leaves the actors vague. It also suggests that in the act, and only then and there, something *is*—being enacted. Both suggestions fit in fine with the praxiography that I try to engage in here.

Thus, an ethnographer/praxiographer out to investigate diseases never isolates these from the practices in which they are, what one may call, *enacted*. She stubbornly takes notice of the techniques that make things visible, audible, tangible, knowable. She may talk bodies—but she never forgets about microscopes. This turns the distance from the outpatient clinic (which, in hospital Z, is located on the first floor of wing F) to the department of pathology (on the fourth floor of wing D) into one that is very long indeed. An unbridgeable distance, or so it seems. For the techniques that make atherosclerosis visible, audible, tangible, and knowable in these two places exclude each another.

We walk to the fridge. The pathology resident takes out a plastic bag with a label attached to it. Inside it there's a foot with twenty-eight centimeters of leg. It was amputated the previous day and routinely sent to the pathology department for inspection. Could the plane of resection, the skin, and the vessels please be prepared and assessed under a microscope? While he carries the amputated lower leg to a table, the resident puts his hand on the place where one might expect the dorsal foot artery. "Hah, nice pulsations," he says provocatively. And then he looks at me and adds: "Ain't I horrible?"

at stake. Sure, they depend on one another. The many dichotomies that infest the modern philosophical tradition are all interrelated. And yet there are also endless varieties and incongruities between them. When it comes to an investigation of disease by ethnographic means, it is important to stress the double character of the subject/object divide. The subtext making relations to the literature throughout this chapter aims to show just this. That there is, first, a division between subject-humans and objects-nature. And that second, there is a related but different division between actively knowing-subjects and passive objects-that-are-known. Escaping from the first dichotomy involves different moves than getting away from the second.

Subjects/Objects 1

If humans, who can talk, are, because of this ability, to be respected as subjects, while other entities, silently part of nature, may be turned into objects, then the question arises: which of these kinds of entities may scholars hope to publish printed texts about? There is a long-standing differentiation: the social sciences know about humans and their societies, while the natural sciences know about the natural world. A lot of disciplines do not fit into this scheme: geography, architecture, and medicine to name but three. And yet it is persistent. There are various reasons for this. One of these is that many social scientists fear that as soon as the divide is not respected, natural scientific methods will take over. Imperialistically they will reach

In the outpatient clinic, surgeons feel the pulsations of dorsal foot arteries in patients whose legs hurt when they walk. Each time the heart beats, a person's blood is pushed forward through the arteries, and this can be felt on the body's surface (in contrast with flow through the veins, which carry the same blood a lot calmer back to the heart again). In the pathology department the gesture of feeling for pulsations is empty. The arteries of dead limbs do not pulsate. It is a sick joke to feel for them even so.

He's a good informant, this resident, even if he makes sick jokes. Or, he's a good informant because he makes sick jokes. Jokes that may have a psychological function: they may facilitate this young man's entrance in the esoteric world of pathology, where, unlike most other places, cold human lower legs are things one may take out of a fridge and walk around with. But the joke quoted here also contains ethnographic information. It enlarges the fact that the requirements for enacting disease in a clinical way are no longer met once a patient is dead. However skilled a novice doctor may be in feeling pulsations, this is not going to help him when it comes to diagnosing the vessels of an amputated lower leg.

In the department of pathology, no pulsations can be felt and no interview questions can be asked. Does this leg hurt? Even if there were a patient present who might want to answer such a question, it wouldn't make sense. Either a leg

everywhere and human subjects, instead of being listened to, will get objectified. (For a debate about this question, see, for example, the debate between Collins and Yearly and Latour and Callon, in Pickering 1991.) But not respecting the divide also opens another possibility, one that is hardly ever mentioned: it might also be that the social sciences have methods that are capable of reaching out, of going everywhere—even if they can't do everything. Indeed, or so I will argue here, methods like this exist. One of these is a sociological tradition designed for the study of human subjects. If pulled and pushed a bit, it may be broadened to encompass subject/objects of all kinds.

In order to make this claim, I begin by taking you back to another outdated text. In 1959, Goffman borrowed the language of the theater in order to talk about

human subjects. When people present themselves to each other, Goffman said, they present not so much *themselves* but *a self*, a persona, a mask. They act as if they were on a stage. They *perform*. In everyday life people present themselves to each other. And while acting, they treat the other people present as both their coactors and the audience of their play (Goffman [1959] 1971). With his suggestion that we might investigate performances, Goffman opened up the possibility of a sociology of the individual. He launched a study of social selves. In shops, factories, churches, pubs, schools, hospitals, and other settings where sociologists may venture and observe what happens, identity is not expressed: it is performed.

Goffman's sociology was designed as a supplement to a specific kind of psychology. Not a static character typifica-

is part of the living body of a patient who is able to talk about it, or a leg is cut off. And however much its absence may hurt, the absent leg no longer hurts itself. In a department of pathology, several crucial requirements for enacting atherosclerosis in a clinical mode are lacking. In the outpatient clinic, it is the other way around. There the techniques of pathology are out of place. They cannot be applied. Making a cross section of an artery is fine—if one has an artery. But nobody is going to cut an artery out of a living body in order to find out how bad it is. Doing so would cause a problem bigger than the one needing a solution. Is it thick, the intima of the femoral artery of that patient, who's sitting on his chair so sadly? It may well be. Who knows? Nobody does. As long as the patient's skin is left intact, no head will bend over a microscope and observe cross sections of the patient's vessels.

The practices of enacting clinical atherosclerosis and pathological atherosclerosis *exclude* one another. The first requires a patient who complains about pain in his legs. And the second requires a cross section of an artery visible under the microscope. These exigencies are incompatible, at least: they cannot be realized simultaneously. This is not a question of words that prove difficult to translate from one department to the other. Surgeons and pathologists who talk with one another tend to understand each other very well. It is not a question of looking from different perspectives either. Surgeons know how to look through microscopes and pathologists have learned how to talk to living patients. The incompatibility is a practical matter. It is a matter of patients who speak as against

tion, nor some behaviorist variant that has only room for input-output correlations, but a dynamic psychology in which, after a developmental process, adults have real selves deep down, back stage. In *The Social Presentation of the Self in Everyday Life,* Goffman left this backstage identity on one side as a topic to be studied by psychologists. The sociological object was framed as something differently. The identity people *perform* is not deep, it is a *mere* performance. Due to his sociological training, or so he claimed, Goffman had enough distance so he could always see the curtains. But to the players and everyday, nonsociological observers, the gap between performance and reality often goes unnoticed. They may be carried away by the play. In Goffman's words: "At one extreme, one finds that the performer can be fully taken by his own act; he can be sincerely convinced that the impression of reality which he stages is the real reality. When his audience is also convinced in this way about the show he puts on—and this seems to be the typical case—then for the moment at least, only the sociologist or the socially disgruntled will have any doubts about the 'realness' of what is presented. At the other extreme, we find that the performer may not be taken in at all by his own routine. This possibility is understandable, since no one is in quite as good an observational position to see through

body parts that are sectioned. Of talking about pain as against estimating the size of cells. Of asking questions as against preparing slides. In the outpatient clinic and in the department of pathology, atherosclerosis is *done* differently.

Founding or Following

There is a certain economy in isolating objects from the practices in which they are enacted. When the intricacies of its enactment are bracketed, the body becomes established as an independent entity. A reality all by itself. Alone and self-sufficient. This makes it possible to relate the pain articulated in the consulting room with the thickened intima visible under a microscope. It is possible. Forget about "articulated in the consulting room" and "visible under a microscope" and pretend that both practices share a single, common object. They have as their *referent* a single disease, residing *inside* the body. In its leg arteries, to be precise. It surfaces in symptoms, the patient's complaints among them. And it is unveiled when the vessels are finally put under the microscope.

It often happens. The practicalities of enacting disease are bracketed. Atherosclerosis is taken to be one disease. The patient's pain is among the *symptoms* that surface, and the thickened vessel walls are called the *underlying reality* of the disease. This layered image turns pathology into a crucial discipline, for it unveils the underlying reality of disease. Pathology is, indeed, called the *foundation* of modern medicine by many analysts for that very reason. Some simply

the act as the person who puts it on" (28). But while the psychological "realness" of the identity on stage might be doubted (by the sociologist, the socially disgruntled, and the person who puts it on), the social consequences of the publicly displayed role are impressive even so. The identity people perform in public, on stage, is the one others react to and is thus the one that is socially effective. It is, therefore, an important object of sociological study.

Again, the outdated text I relate to here has been covered up later on by many other texts (written by Goffman himself as well as by other authors) that tell more or less different stories about identity and/or performance. Instead of digging out that history in detail here, I will make a big

jump (over lots of intricate details) to two texts of a few decades later. One articulates clearly the idea that, somewhere in-between, along the way, the curtains have vanished. The other broadens the study of performances from human identities to entities of heterogeneous kinds.

Somewhere between the fifties and the eighties, psychology lost its power to study the real reality of individuals. Sociology, when observing what individuals do in public, *on stage*, no longer feels that there is anything deep that it is missing out on. In terms of the stage metaphor, one could say that there are *only* stages these days. The curtains and the dressing rooms have gone. Sociologists take sociological reality at face value. The "mere" has dis-

assert this. Others see it as a reason for criticism: what kind of medicine is this, that wants to heal living patients but is based on the knowledge of dead bodies?

However, if one doesn't bracket the specificities of enacting reality the picture changes drastically. If one doesn't stay within the confinements of the body, but follows the various practices in which atherosclerosis is enacted throughout the hospital, the topography of the relation between pathology and clinic appears to be completely different. In hospital practice, thickened vessel walls do not *underlay* legs that hurt. They come, instead, *after* them. And, moreover, they only do so for a small proportion of patients. In practice, thickened vessel walls are only revealed in those patients whose legs have been amputated or in those who have been operated on and from whose bodies small parts are sent up to floor 4 wing D to be put under the microscope. In practice, if pathology has anything to do with atherosclerosis at all it is not as a foundation, but as an afterthought.

The pathology resident carries the amputated foot-with-leg that he just took out of the refrigerator to a table. He measures the length of the leg: twenty-eight centimeters. Makes a note of that. Then he takes a dissection knife out of a drawer. He cuts two small pieces

appeared from the performance. "My argument is that there need not be a 'doer behind the deed,' but that the 'doer' is variably constructed in and through the deed," writes Judith Butler while talking about doing gender identity (1990, 142). The opposition between surface appearance and deep reality has disappeared. And people's identities do not precede their performances, but are constituted in and through them. Identity depends on what happens on stage: but then psychology is either wiped away, or turned into another branch of sociology.

The specific identity that Butler is concerned about is that of *gender*. Turning this into a topic for sociological investigation is a way to push aside another tradition that claimed to know about it: psychoanalysis. Psychoanalytic stories say that early on in people's lives their identity is not yet fixed: it may still take various forms. But somewhere before the age of four one becomes either a woman or a man. This, then, is what Butler challenges. "What is signified as an identity is not signified at a given point in time after which it is simply there as an inert piece of entitative language" (144). Identity, Butler tells, is not given but practiced. The *pervasive and mundane acts* in which this is done make people what they are. These acts deserve to be taken seriously both in their stubbornness and in their volatility.

But how to study the acts in which people *do* their selves? How can one avoid being *taken in* by the face-value reality of what happens if one no longer frames the stage in the way theaters do, with curtains, but as if one is making a documentary film with a hand camera that may be carried everywhere? Goffman had his scholarly distance to rely on when he studied performances, a distance that made

of tissue from the plane of resection, puts them in plastic containers, and numbers these. He scribbles the numbers in his notebook next to a rough drawing, indicating with arrows where each specimen was taken from. He does the same with a few pieces of skin. Then he starts to look for the arteries. It's not easy to find these now that they do not pulsate. But finally he succeeds. He cuts several pieces of each and puts them in containers as well. The containers have holes. They are all dropped in a small bucket that is filled with a fluid that will prevent their disintegration. The next day technicians will turn the preserved pieces of tissue into slides. And in a few days' time the resident and I will be bend over the microscope and see arteries with impressively thick intimas: atherosclerosis. We'll also inspect the cells of the plane of resection. They look all right: not gangrenous. And the skin cells indeed show the signs of long and severe oxygen deprivation. The resident writes this down and takes his notes to his supervisor.

Pathology has the final word in cases of amputation. While the patient is recovering in a hospital bed and learning to live with an incomplete leg, pathologists decide whether the operation was justified and properly executed. Pathologists may also make judgments about the walls of small pieces of arteries that are cut out of poorly functioning circulatory systems in the course of less drastic operations. They may judge all kinds of arteries once they do not function

him aware of the curtains, but what does Butler have, so many years on? Friction. Differences. Contradictions. "The injunction *to be* a given gender produces necessary failures, a variety of incoherent configurations that in their multiplicity exceed and defy the injunction by which they are generated" (Butler 1990, 145). Clashes and transgressions make diverging rules and regulations visible. Because *doing* a woman is not the same thing in a supermarket as it is in the classroom, because *staging* a man in bed is quite different from staging a man at a professional meeting, it is possible to investigate what it is to perform this, that, or the other gender. Instead of distance, now, here, it is *contrast* that makes it possible to be a good observer.

Human subjects can be studied in this way: by investigating their contrasting identities as these are performed in a variety of sites and situations. But what about the entities of the natural world, the objects? The investigation of gender identity in terms of performance begins by diminishing the importance of a few natural objects. The vagina for instance. This organ is no longer capable, all by itself, of turning someone into a woman. A lot more is required to *do* womanhood: specific styles of talking, ways of walking, dressing, addressing. A womanly way of screaming, raging, smiling, eating, soothing, loving. If gender is not fixed and physical but viscous and performed, the body's sexual organs are not enough to mark it.

But then again. Performing identities is not a question of ideas and imaginations devoid of materiality either. A lot of *things* are involved. Black ties and yellow dresses. Bags and glasses. Shoes and desks and

any longer, once the blood has stopped flowing through them. But they never answer the question "what to do?" that drives the enactment of atherosclerosis in the clinic. In the daily hospital dealings with patients with atherosclerosis, pathology is not foundational, because it cannot found action. However basic its truth, pathology cannot get to know what vascular surgeons want to know when they make decisions about treatment. Should this patient, Mr. or Mrs. So-and-So, be operated, and, if so, where, and how? Pathology remains silent on these questions.

To the pathology resident it is frustrating. He expected this specialism to be basic and thus to have all knowledge, an overview. But often it cannot even answer simple questions. As he puts it: "I'll never be able to diagnose the state of an artery properly. Never. Not even if I have an entire vessel. In a living patient this is ridiculous of course. But I couldn't even do it in a corpse. For what do you want to know? You want to know the location and extent of the stenosis. That implies that you'd have to make a slide every, say every three centimeters. Or maybe five. Just imagine: over the entire length of a lower leg, an upper leg, an aorta. How many slides is that? Imagine me cutting all the pieces. The technicians slicing them, coloring them, making slides. And then I'd have to assess these carefully, one by one. It wouldn't be enough to say that the wall is thick. How thick is the wall? How much of the original lumen is left? I'd have to take into account that I look at a lumen that is no longer functioning. It would take ages. It's time consuming so it's far too expensive. And because there are all these artifacts of death, it's not even certain either. It can't be done."

chairs and razors. And among the stage props is the physical body. A vagina or a penis need not cause gender identity from the inside to be relevant in staging oneself as a woman or a man. The extent to which they are relevant depends on the scene. Out in the streets one does not need a penis to perform masculinity. But in communal showers at the swimming pool, it helps a lot. So there they are, the genitals: on stage.

But where are they—where in the literature? Not in Butler's book. Butler is a philosopher who says that it is important to study the pervasive and mundane acts by which gender identity is performed. But she doesn't actually engage in such a study. Others do. Stefan Hirschauer, for instance. As a sociologist he made an investigation of the performance of gender identity (Hirschauer 1993). His point of entrance is an ethnographic study of a German treatment program for transsexuals. Transsexuality, or so Hirschauer states (following Garfinkel), can teach the sociologist a lot about what it is to perform a gender because transsexuals pass from one side of the divide to the other. What is involved in passing? The law, the job market, family relations. And, to be sure, the body. The body is, cannot but be, restyled by the person who is, or tries to be, the "other" gender, the one that his/her genitals do not denote. Length of the hair, length of pace, way of sitting, they are all adapted.

So the transsexual body is part of

In practice, the different ways in which atherosclerosis is enacted do not align. Opening up a leg in order to find out whether its arteries are bad isn't done because taking out a part of an artery for diagnostic reasons would be an intervention as big as a therapeutic one. A biopsy of just a little part of the artery, moreover, wouldn't show *where* it is bad: in the groin, the knee, the ankles? The resident's thought experiment, in which he gives himself an entire vessel to diagnose, shows that even if that impossible condition for his work were met, he would not be of great help to vascular surgeons. Even then he would not gather the kind of information that treating surgeons want in addition to their clinical diagnosis: the location and quantification of a patient's atherosclerosis.

In the process of diagnosing atherosclerosis, the knowledge on which action may be based doesn't come from the pathology department. This is not an accidental division of tasks. The knowledge required could simply never be assembled using the techniques of pathology. What about the clinic? In hospital practice, the clinical way of enacting atherosclerosis is more important. This is not to say that the clinic, in its turn, is foundational. The appropriate term here is another one. The reality enacted in the clinic comes before all others. It is the *beginning of* and the *condition for* everything else. This becomes particularly apparent when patients fail to comply with the unwritten rules of the doctor-patient interview, when patients seem to expect that their complaints and their experiences, their stories, are of no importance to the doctor.

I sit in with the angiologist, an internist specialized in vascular diseases. In the course of the morning, he sees patients with claudication, but also patients who have vascular problems other than atherosclerosis. There are, moreover, patients whom the general practitioner

staging a new gender identity. To that end it is restyled, and not only by the transsexual, but also by medical professionals. Hirschauer's study goes into this medical restyling in a lot of detail. It comes after the psychiatrist has accepted the person's claim to being the other gender. Then this false body is first diagnosed as endocrinologically normal, in order for it to get hormones that make it as normal as possible again, but this time according to the other normal values. Subsequently it is operated on: its genitals are heroically resculptured. Vagina is turned into penis or vice versa. Without those physical interventions, transsexuals, or so they say, have trouble performing the other gender. They need a body with the "right" sex to be able to have a coherent identity. Bodies thus do not oppose social performances, but are a part of them. Performances are not only social, but material as well. So there they are, the objects. They take part in the way people stage their identities. But once objects are on stage we can investigate *their* identities, too. This is what Hirschauer does and also what happens in the present book: here objects are investigated as if they were on stage. What is studied here are the identities an ob-

couldn't diagnose. They are likely to have internal problems, but of what kind? This makes the interview questions more open than they tend to be in the vascular surgeons' clinics. Not: do your legs hurt? But: what can I do for you? Or: what's your problem? Mrs. Vengar comes for the first time, suffering visibly. The angiologist looks up at her from his papers. "Well, what is troubling you?" Mrs. Vengar shakes her head, slowly. And then she says: "I don't know doctor, I don't know what it is that troubles me. That's what I come to see you for. Because I don't know."

An answer like that leaves a doctor in an outpatient clinic with empty hands. He's been there before. It is an awkward place to be. He has to get her to talk. A doctor cannot hope to guess where to begin with his further diagnostic work without some significant answer to his interview questions.

There are provinces of medicine in which the clinic doesn't take the lead. In cancers, the microscopic images of the pathologists are likely to overrule clinical stories once they are available. Biopsies are taken out of lungs, livers, breasts, and many other organs in order to inspect small slices of tissue under the microscope. The pathologist gives the diagnosis. For some diseases this is even done before patients have complaints about which they might speak. In the Netherlands and in various other countries, Pap tests are offered to women of designated ages in order to detect early stages of cancer of the cervix. So, pathology is of primary importance in medical dealings with cancers.

However, in large parts of medicine, and certainly in the hospital's dealing

ject may have when staged, handled, performed.

In the literature there has been a lot of discussion about the term *performance*—a term that does not only resonate the stage but also success after difficult work and the practical effects of words being spoken. I do not want these resonances, nor do I want this text to be burdened with discussions that it seeks no part in. But if one doesn't want to be a part of, let alone be played out in, controversies raging in *the literature*, if one doesn't want one's texts to be grinded between concerns that aren't one's own, then what can be done? It may be helpful to avoid the buzzword. To look for another term. A word that is still relatively innocent, one that resonates

with fewer agendas. I have found one. And, even if I have been using the term *performance* elsewhere in the past, I have carefully banned it from the present text. I use another verb instead, *enact,* for which I give no references, precisely because I would like you to read it in as fresh a way as possible. In practice, objects are enacted.

Talking about the enactment of objects builds on and is a shift away from another way of talking about objects, one in which the term *construction* has a prominent place. From the late seventies to the early nineties objects were thematized in ways analogous to psychodynamic investigations of subjects. During that period, the term *construction* was widely used, and the term *making* also appeared frequently (just

with atherosclerosis of the leg arteries, pathology doesn't have such a strong position. Instead, the reality of the outpatient clinic comes first. This doesn't mean that the patient's story is always taken at face value. But it certainly implies that the patient's story either opens up or forecloses further moves along the diagnostic and therapeutic track of atherosclerosis.

The vascular surgeon says to Mr. Zender, a man in his early forties: "Now, tell me, what's your job?" Mr. Zender answers with the name of a job I've never heard before. Neither has the surgeon, for he says: "Well, I don't know what that is, but please don't explain it to me, just tell me: do you have to walk a lot?" "No," says the patient, "it's mostly sitting. But recently, with this pain in my legs, I find myself looking for an excuse to walk. Go to the second floor. That kind of thing." "So, do you. What if you sit down at home?" "You see doctor, as long as I do things, it's all right. But like, if we've done the washing up, children to bed, sit on the couch in front of the television, then it starts hurting." The surgeon summons Mr. Zender to the examination table. And says meanwhile: "I'll just have a look to reassure you. So that you won't say I didn't even examine you. But let me tell you one thing. You may have pain in your legs all right. But there's nothing wrong with your leg arteries."

In the vascular surgery outpatient clinic, it is clear and distinct. This story isn't about atherosclerosis. In severe cases, patients with atherosclerosis may

two examples: Edward Yoxen, "Constructing Genetic Diseases," 1982; and Cecil Helman, "Psyche, Soma and Society: The Social Construction of Psychosomatic Disorders," 1988). The term *construction* was used to get across the view that objects have no fixed and given identities, but gradually come into being. During their unstable childhoods their identities tend to be highly contested, volatile, open to transformation. But once they have grown up objects are taken to be stabilized.

One of the pivotal texts (everybody relates to it, I might as well) is *Laboratory Life* (Latour and Woolgar 1979). It tries to get away from a place where reality is supposed to have fixed traits. "Scientific activity is not 'about nature,' it is a fierce fight to *construct* nature. The *laboratory* is the workplace and the set of productive forces, which makes construction possible. Every time a statement stabilizes, it is reintroduced into the laboratory (in the guise of a machine, inscription device, skill, routine, prejudice, deduction, program, and so on), and it is used to increase the difference between statements. The cost of challenging the reified statement is impossibly high. Reality is secreted" (243). The image is beautiful: just as glands secrete hormones, laboratories secrete reality. And yet in the nineties the idea that it is always expensive to change the identities of objects has started to lose ground. By that time we may read that "matter isn't as solid and durable as it sometimes appears. And if it does hold together? Well, this is an astonishing achievement" (Law and Mol 1995, 291).

(But who put this into the literature, to draw it out again now, in a quote? Hmm, I was one of the authors doing so. Does

have pain even when resting, but then their legs will hurt a lot more when they walk. And if someone looks for an occasion to move his aching legs when resting, he may be in trouble, but such trouble cannot be eased by the vascular surgeon. The surgeon shrugs his shoulders when asked where the pain may come from, says he doesn't know, and refers the patient back to the general practitioner. It is only when a patient articulates the complaints specific to atherosclerosis that vascular surgeons start to do a physical examination with the expectation of finding the disease they are feeling for.

The Objects

When the practicalities of enacting disease are stressed, not bracketed, it becomes clear that pathology does not play a foundational role in the diagnosis of patients with atherosclerosis in their leg arteries. If it plays a role at all, it is as an afterthought. A well-aimed clinical interview is far more important: it takes the lead. But what follows from this? One might attach a "merely pragmatic" significance to it. One might unbracket practicalities, admit they exist, even pay attention to them, and yet still see them as a subordinate matter. Something to do with the state of the art, limits to the possibility of knowing, but not the reality of the body. Someone arguing in this way would say that even if pathology isn't the foundation of medical practice, thick vessel walls are still causing complaints.

quoting one's own earlier words still work to situate a later text, or does relating to the literature only make sense if *the literature* and *the author* are two separate, different, bounded and exclusive entities? It's up to you. Does it work here?)

In a variety of sites in the nineties the idea that objects might not just gradually acquire an identity that they then hold on to has been pushed aside, or complemented, by this new idea. That maintaining the identity of objects requires a continuing effort. That over time they may change. If I claim that this is *in the literature,* why then not relate to Charis Cussins here? She makes the objects dance, and her title alone is telling enough for what I try to convey: that there is an ongoing "onto-

logical choreography" (Cussins 1996). The present book is one of the products, symptoms, or elements of the process of *decentering the object* (as John Law calls it in Law 2002). It does not simply grant objects a contested and accidental history (that they acquired a while ago, with the notion of, and the stories about their *construction*) but gives them a complex present, too, a present in which their identities are fragile and may differ between sites. It does so by deploying sociological, and more specifically ethnographic, methods of study. By describing the various performances — or enactments — of the objects' identities on stage.

Thus, the remarkable shift has been made: a social scientific way of working has

The question is: are they? Beware. I won't answer this question with a straightforward "yes" or "no." For when they move beyond the disease/illness distinction, ethnographers may talk of bodies—but not of isolated bodies. So I won't speak about the relation between vessel walls and complaints *inside* the body here. I'll stubbornly stick to studying "reality enacted" and will approach the question ethnographically yet again.

The pathology resident takes his notes to his supervisor. "I've checked everything," he says, "the cells in the plane of resection were fine, so they've done their amputation high enough. The skin cells showed signs of long and severe lack of oxygen. They were in complete shambles. And all my cross sections were of very sick vessels. Thick intimas, hardly any lumen left." The supervisor takes the notes. Wants to know a few further details. Comments on the slightly off use of a technical term. And then says: "Okay. I'd better have a last look at your slides and sign the report. They can be happy. They've been approved."

Pathology may not be the foundation of all medical action but in cases such as these it judges what was done. The surgeons did an amputation because even when at rest the patient was in agony, his skin was in a very poor condition, and there was no possibility of improving his circulation. His lower leg was cut off. This specificity allows pathology to be practiced. It comes after the clinic, but only shortly after it. Just a few days. Thus, their objects can be compared. The pain of the clinic and the thick intimas of the pathology department are mapped

come to extend itself to encompass the physicalities whose study used to be the prerogative of the natural sciences. The dividing line between human subjects and natural objects has been breached—but not in a way such that physics can take over the world, or that genetics is allowed to explain us all. The (serious) game played here makes a move that is the other way around: like (human) subjects, (natural) objects are framed as parts of events that occur and plays that are staged. If an objects is real this is because it is part of a practice. It is a reality *enacted*.

Subjects/Objects 2
Since the time sociology invented "illness" as an object of study in its own right, it has tried to add knowledge of the illnesses

people live with to that of the diseases that plague their bodies. Philosophers tend to frame a similar concern in terms of minds and bodies. The hope keeps coming back. Sociopsychological subjects and natural objects should both be attended to. Here's a quote from the early eighties: "We are now faced with the necessity and the challenge to broaden the approach to disease to include the psychosocial without sacrificing the enormous advantages of the biomedical approach" (Engel 1981, 594).

Addition is advocated over and over again: psychosocial insights must be added to biomedical facts. But this is not the only way of pressing medicine to overcome its neglect of human subjectivity. There's another one as well. It appears, for instance, in the answer Mark Sulli-

onto one another. They are both impressively severe. It turns out that there is atherosclerosis in the one just as much as there was in the other. The objects of clinic and pathology *coincide*.

In order to find out whether the objects of the clinic and pathology indeed coincide, they must be related. When does this happen? When are clinical and pathological atherosclerosis related? In the process of deciding about the treatment of a patient who has pain on walking, they are not. But as soon as a piece of vessel is available, a link can be made. Then it is possible to make a cross section and ask if the thickness of the vessel wall is as impressive as the complaints that were uttered just a little earlier in the clinic. This may be the case. The objects of clinic and pathology may coincide. Sometimes, however, they do not.

The pathologist: "You, since you're so interested in atherosclerosis, you should have been here last week. We had this patient, a woman in her seventies. She had renal problems. Severe ones, too. So she was admitted. And the next day she died. Paff, from one moment to the next. The nephrologists were aghast, and so, of course, was her family. So we were asked to do an obduction. It was unbelievable. Her entire vascular system was atherosclerotic. One of her renal arteries was closed off, the other almost. It was a wonder her kidneys still did anything at all. It was hard to see where they got their blood from. And it was more or less the same for every other artery we took out: they were all calcified. Carotids, coronary ar-

van gives to the question he uses as a title: "In what sense is contemporary medicine dualistic?" (Sullivan 1986). Sullivan argues that instead of adding the patient's subjectivity to medicine's *objects* of inquiry, it should be approached quite differently: as a knowing instance, a *subject* of knowledge. Contemporary medicine, says Sullivan, inherited its dualism not from Descartes, but from Bichat. Bichat stood at the cradle of modern pathology. His work marks the moment in the early nineteenth century when pathology came to take its foundational place in medicine— says Sullivan. "For Bichat, the medical subject and the medical object were not two different substances within the same individual, but two different individuals: one alive and one dead. Knower and known are epistemologically distinguished with

the physician assuming the position of the knower and the patient/corpse the position of the known" (344).

Where the dissection room is turned into the place where truth about diseases may be spoken, the patient is silenced. "Here, the *activity* of self-interpretation or self-knowledge is eliminated from the body rather than the entity of mental substance. The body known and healed by modern medicine is not self-aware" (344). The verdict is stretched out from Bichat's writings to "modern medicine," for this has not left the episteme, the mode of knowing, that rose with the birth of the clinic. (If I left it at that, the last sentence would contain an *implicit* reference to the literature, to a book, that I think I had better make explicit. It is a book that inspired so many later writings about medicine, Sullivan's

teries, iliac arteries: everything. Thick intimas, small lumens. And she'd never complained. Nothing. No chest pain, no claudication, nothing. We phoned her general practitioner just to check it. He said she'd been visiting him for coughs and things. High blood pressure. But not with any complaint that made him think of atherosclerosis."

The pathologist remembers this patient well because her condition surprised him. Pathologists expect bad vessel walls to cause complaints. But for one reason or another this expectation isn't always fulfilled. The pathologist quoted here rightly takes this to be a phenomenon of interest to the observer.

If a relation between the atherosclerosis of pathology and the atherosclerosis of the clinic is made, in practice, their objects may happen to coincide. But this is not a law of nature. It may also happen that a patient who never complained turns out to be severely atherosclerotic at the postmortem. In such a case, the objects enacted in the clinic and in the pathology department don't map. They *clash*. One atherosclerosis is severe while the other isn't. One atherosclerosis might have been a reason for treatment while nobody ever worried about the other. In such instances the objects of pathology and clinic cannot be aspects of the same entity: their natures are simply not the same. They are different objects.

Explanations will be sought. Did the patient suffer from pain but never report it? Did she always sit and avoid walking? Had her condition developed so slowly that her metabolism had adapted itself? Sometimes it is possible to find

analysis among them: Michel Foucault's study *The Birth of the Clinic* [1973].) Sullivan states that we haven't left the modern era. All knowledge assembled in the hospital still refers to a body in which surfacing symptoms point toward the underlying deviance of tissues. If doctors hear a complaint in the clinic, they try to link it to a deviance that would be visible when the patient's tissues were inspected in the department of pathology. This is only possible when the body is a corpse—or when at least the tissues themselves, cut out of a living body, are definitely dead.

Here, then, we have the second subject/object divide: a distinction between knowing subjects and objects known. It does not run parallel to the first. Since the birth of the human sciences, human subjects (whether carefully separated out from so-called natural objects or not) can have two positions in relation to knowledge: that of subject and that of object. How to escape from this divide? "Any attempt to redress the shortcomings of the clinico-pathological approach to patients must address itself not to some vague reintegration of mind into the medical body. It should rather concern itself with a reappropriation of patients' capacity for self-knowledge and self-interpretation into our definition of disease. Put as succinctly as possible: the meaning of the disability for the patient must be incorporated into the very definition of that disability as disease" (Sullivan 1986, 346).

an explanation for the difference between the objects of pathology and clinic. But even if clashes between different "atheroscleroses" can be explained, they cannot be explained away. They have a consequence. Inevitably. A practical consequence. If two objects that go under the same name clash, in practice one of them will be privileged over the other.

The vascular surgeon: "Oh no. No, we don't dream of it. We'll never go out into the population to find all the bad arteries around. For if we did, and if we then offered an operation to all those patients it would simply cost a fortune. And, more important, we would create far too many victims. If people have severe complaints, you may improve their condition. But if they have no complaints, or few, they don't have enough to gain. While they still run risks. Sometimes an operation makes things worse. Or people die. So you're not going to cut if their lives won't be improved by it."

In as far as the object of pathology clashes with that of the clinic, this is too bad for the thick vessel walls that go undetected. In the current practice of dealing with atherosclerotic leg arteries, the clinical way of working wins. Nobody in hospital Z is going to sift out all the people in the population of the region who may have thick intimas and small lumens and yet do not get treated surgically. The detection of atherosclerosis of the leg arteries is organized along clinical lines. You only ever become a vascular patient if you visit a doctor and say that you have pain on walking.

Thus, the fact that pathology isn't the foundation of all medical practice, but that the clinic takes the lead when it comes to the diagnosis and detection of this disease, is not a merely pragmatic matter. It touches reality all right. It doesn't

Something complicated is happening here. When critics (like Sullivan) say over and over again that medicine silences the objects of its knowledge, the irrelevance of what patients have to say is restated as many times as a fact. Thus, the fact is strengthened. There might be better ways of escaping. (Another reference to Foucault is suitable here: he masterfully argued for noncritical strategies for escaping dominant ways of thinking, and he engaged in them himself. See, for example, his claim that criticizing "sexual oppression" is not a revolutionary act, but just another expression of the configuration of sexuality that we have been living with since the late nineteenth century, in which sexuality is an urge tamed and domesticated as if it were a wild animal [Foucault 1981].) It might then be a good way to escape from a medicine founded on pathology to wonder whether, in practice, medicine *is* indeed founded on pathology. This implies that instead of *criticizing* pathology's foundational role, we raise questions about it, we *doubt* it. That we don't go with the textbook versions of medical knowledge, but analyze, instead, what happens in medical practices. Sullivan tries to challenge Bichat's definitions of disease by

make complaints *more real* than the size of vessel walls. But it does turn them into what will *count as the reality* in a particular site. Not under a microscope, this time, but in the organization of the health care system. *Under the microscope* atherosclerosis of the leg arteries may be a thick intima of the vessel wall. *In the organization of the health care system,* however, it is pain. Pain that follows from walking and that nags patients suffering from it enough to make them decide to visit a doctor and ask what can be done about it.

Which Site?

If the practicalities of enacting disease are bracketed, disease is located inside the body. In the legs or in the heart. In the aorta or in the leg arteries. In the groin or near the knees. Anatomy helps to say *where* things are wrong: it is an important topographical language for talking about bodies. It is not only used by pathologists when they do a dissection, but in the consulting room as well. "Where does it hurt?" physicians tend to ask their patients. Most patients visiting hospital Z have learned to answer this question somehow. They designate the sites of their bodies that hurt with a pointing finger. The doctor may translate such answers into anatomical terms and write down "lower abdomen left" or "posterior crural region right" in the patient's file.

However, the ethnographer who persistently attends to practicalities needs another topographical language. Or maybe several. If reality is enacted differently from one site to another, the question about *where* these sites are cannot be answered by a finger pointing at the regions of a body. The practicalities of

adding knowing patients to known bodies. I prefer to try to challenge Bichat's definitions of disease by doubting the assumptions of the relation between knowledge and practice that come with it. Is pathology indeed foundational if we no longer investigate medicine as if there are knowing subjects on the one hand and objects to be known on the other?

I am talking about Sullivan in order to show what, in relation to the literature, I am doing when I investigate the place of pathology in the diagnosis of atherosclerosis and contrast this with what is said and done in the outpatient clinic. I am trying to find whether, indeed, patients are silenced

and pathology is foundational. And instead of studying these topics by teasing out what doctors know or what happens to patient's self-knowledge, I have analyzed the knowledge incorporated in practices. The knowledge incorporated in practices does not reside in subjects alone, but also in buildings, knives, dyes, desks. And in technologies like patient records — as David Armstrong tells us in an article that wonderfully shows how the material organization of medical practice shapes the reality of disease. Armstrong's claim is that pathology is no longer foundational since in practice disease is no longer projected onto the body's various layers

medicine are to be found in other places. But which? So far I've given a few in-dications. I've said that my observations were conducted in a university hospital in a medium-sized town in the Netherlands, hospital Z. That's a specific place. But I've also differentiated between two sites in this hospital: the department of pathology and the outpatient clinic. Atherosclerosis, however, is enacted in other sites and other kinds of places as well.

We've seen that the detection of atherosclerosis does not proceed through screening the population but by waiting for patients. *Where* does this statement hold? In hospital Z, for sure, but also in the Netherlands as a whole. And the area is larger still. This policy of waiting characterizes the detection of atherosclero-sis in all Western countries. Or in all countries where cosmopolitan, allopathic medicine is practiced. But in an area this large, it will surely be possible to find exceptions. As indeed it is. There are even exceptions to what I've said so far within the close confines of hospital Z.

The internist has been working for three years in hospital Z now. It's the second morning I sit on my small stool behind him. "Oh, gosh," he sighs when one patient has just left and he sees the file of the next. He explains his sigh to me. "The next man is someone I inherited from my predecessor. It's a perfectly healthy but slightly neurotic professional in his fifties who wants me to turn him inside out. He's particularly anxious about getting atheroscle-rosis. I don't believe I can find anything that we might be able to do something about. If he were developing atherosclerosis, all I could do is advise him to move a lot, do sport, eat

(symptoms surfacing and a lesion in the tissues underneath). Instead, disease has become a process in time. "Before records, every patient, every 'contact,' was a sin-gular event; there may have been a 'past history' in the consultation and indeed the doctor might have remembered a signifi-cant past occurrence but past and present were different domains of experience. With the record card, however, which marked the temporal relationship of events, time became concatenated. Clinical problems were not simply located in a specific and immediate lesion but in a biography in which the past informed and pervaded the present" (Armstrong 1988, 217).

Armstrong presents it as a historical se-quence in which one configuration follows on the next, which isn't quite what I am after, if only because pathology hasn't vanished. It somehow coexists with the medical record. But the interesting bit is that Armstrong mentions the record, that he takes it to be important, that he won-ders what such technologies may do with the lived reality of a disease. In Arm-strong's article, knowledge does not re-side in minds. Instead, materials are ac-tively engaged in the enactment of reality. Records, buildings, knives. And then, why not corpses, too.

A corpse is spread out on the metal table of the dissection room in the department of pathology. It is about to be dissected. But however mute, this corpse is active. It tells that someone's life has ended. It tells

wisely, and refrain from smoking. But I can tell him that before I do a battery of tests on him. I do what he wants even so. I tried to talk about it with him last time and the time before that and I've given up. He can have some tests, if that's what he needs to feel safe."

Here a person who has no complaints at all is going to be submitted to diagnostic tests. A pathological inspection of his vessel walls won't be among these, and yet the clinic doesn't take the lead here either. The "patient" about to enter the consulting room has no complaints. His legs do not hurt on walking. So the area where the clinic takes the lead in the detection and diagnosis of atherosclerosis is not fixed. It is a very large site. The area of distribution of "allopathic medicine" is huge. But even in hospital Z it is fairly easy to find exceptions.

If I do not speak about "Western medicine" in this book, nor make claims about other large-scale regions, this is because doing so would skip over too many exceptions. And yet the stories I tell here are not only about what happens in hospital Z. With some changes, shifts, and specific alterations, they might be told, to some extent, by someone else, some other time, about a lot of other hospitals—hospitals in the Netherlands (there is a lot of *Dutchness* in the stories I relate here) but also hospitals anywhere else where there are hospitals. So the area where my stories hold is larger than the one in which they are situated. But it is smaller, too. If I slightly altered the lenses of my ethnographic microscope, or shifted my view sideways a bit, I would tell different stories. The specificities would differ. However, what wouldn't differ is the coexistence of different ways to enact any one disease—the coexistence of different diseases enacted. The fact that there is multiplicity stays the same, in every site, on every scale.

The atherosclerosis enacted in the outpatient clinic contrasts with the thick vessel wall that can be observed through a microscope. But the outpatient clinic

of death—and in a present-day hospital this means that it signals failed treatment. Thus, while a pathologist takes a sharp knife and starts making an incision, the corpse enacts the treating physicians as having failed, as disappointed but also disappointing. As having hit their limits. With a figure of speech one might say that the corpse *knows* the treating doctors failed. I won't use such figures of speech here. Not for knowing corpses. But not for knowing doctors or knowing patients either. This is my claim: that it is a "figure of speech" in both instances. And that it may be a good

methodological strategy to withhold from doctors and patients the subjectivity we are reluctant to grant to corpses in order to analyze embedded knowledges instead.

This, then, may be a way out of the dichotomy between the knowing subject and the objects-that-are-known: to spread the activity of knowing widely. To spread it out over tables, knives, records, microscopes, buildings, and other things or habits in which it is embedded. Instead of talking about subjects *knowing* objects we may then, as a next step, come to talk about *enacting* reality in practice.

is no natural unity. It forms a unity in *contrast* to pathology. When it is approached a little more closely, the clinic appears to be full of contrasts that, in their turn, may be singled out for further investigation. The clinic is not a single site.

A vascular surgeon: "Some of these stories patients tell are so typical, I guess by now you'd recognize them all by yourself. But it's always important to do a physical examination as well. A patient's pain can have many causes. They may even have picked up the story they tell at some party, or from the television. So I carefully feel their pulsations. Inspect their skin. And usually I know from the interview what I will find. But it does happen that a story sounds impressive while the legs are perfectly warm and the foot arteries pulsate happily. I don't like that. I prefer to have a nice, coherent clinical picture."

Doctors don't like it if the atherosclerosis of the interview does not coincide with that of the physical examination. But sometimes it happens. It reveals that the very clinic that I used as a contrast point to pathology is not homogeneous. It does not enact a single object. There are two of them. Two objects. One is enacted through talking, the other through a hands-on investigation. The difference between them may not attract attention as long as the objects they enact coincide, but as soon as they contradict each other it becomes apparent that the clinic is two places. The interview. And the physical examination.

Two? But no, each of these sites can, in its turn, be subdivided into smaller ones. On and on it goes. That conversations between doctors and patients come in endless varieties has often been described. Sociologists have written volumes about this. So let's take the other place, the physical examination.

After I've seen several surgeons at work I look through my notes in order to summarize "the physical examination." But it cannot be done. To be sure these doctors have some gestures in common. They all feel pulsations and the temperature of both feet. But while one always lifts each leg for a while, to see how well the arteries adapt to that, another never does, and yet a third one does so once in a while, for a few patients only.

Blow up a few details of any site and immediately it turns it into many. The ethnographer who counts ways of enacting atherosclerosis, who counts atheroscleroses enacted, won't find an infinite number of variants for the simple reason that there is an end to the number of events that occur in a single hospital — though far earlier there is a limit to her own observation time. But before this limit is reached, the differentiation can go on and on. So what I am trying to relate is not that there are two, five, or seventy variants of atherosclerosis, but that there is multiplicity. That as long as the practicalities of enacting a disease are kept unbracketed, out in the open, the varieties of "atherosclerosis" multiply.

Local Identities

The social science this book mobilizes and tries to contribute to isn't "social" in the conventional sense of the word. This story doesn't tell about people and the relations between them, or institutions and the way they function, or society and whatever it is that generates social order. Instead, this is a story about practices. About events. And one might not even want to call it a story, for there is no smooth ongoing narrative. Instead, I present you with sketches of separate scenes. With snapshots. They are juxtaposed to each other by way of specification or point of contrast. The important roles in these sketches are played by things as well as words, hands as well as eyes, technologies as well as organizational features. Together these heterogeneous ingredients allow me to tell about *atherosclerosis*. Not about the social causes and consequences of the disease, nor about the way patients, doctors, and whoever else involved perceives it. But about atherosclerosis itself. What it *is*.

Those who went through a lot of trouble in order to create a space for the social sciences alongside the natural ones may back away when they come across sentences like "atherosclerosis is. . . ." Or they might get angry. To them such sentences suggest that the domain they conquered with so much effort has been abandoned. To them this careless "is" might make it seem as if the warning that was introduced with so much effort, the warning that we cannot refer unproblematically to objects-out-there but should attend to the activity of referring itself, is being naively thrown overboard. But after the shift from an epistemological to a praxiographic appreciation of reality, telling about what atherosclerosis *is* isn't quite what it used to be. Somewhere along the way the meaning of the word "is" has changed. Dramatically. This is what the change implies: the new "is" is one that is situated. It doesn't say what atherosclerosis is by nature, everywhere. It doesn't say what it is in and of itself, for nothing ever "is" alone. *To be is to be related.* The new talk about what is does not bracket the practicalities involved in enacting reality. It keeps them present.

Thus, atherosclerosis *is* an encroachment of the vessel lumen and a thickening of the vessel wall—in the department of pathology, under the microscope, once a bit of artery has been cut out of a body, sliced, stained, and fixed on a glass slide in order to judge an intervention. But in the outpatient clinic, when surgeons face the question "what to do?" atherosclerosis *is* something else. It is pain that occurs after a certain amount of exercise, pain when walking. It is a poorly nourished skin of one or even two legs, and it is bad pulsations in the dorsal foot artery. The praxiographic "is" is not universal, it is local. It requires a spatial specification. In this ontological genre, a sentence that tells what atherosclerosis is, is to be supplemented with another one that reveals *where* this is the case.

Thus, the trouble taken by social scientists to highlight the importance of representational activities isn't wasted. Instead, it is absorbed into a larger project: there's more work to do, if only because *enacting* is not a question of setting up proper references alone. The enactment of atherosclerosis as an enlarged intima of the vessel wall involves the representational art of making drawings and writing things down, the art of photography and that of printing. But it *also* is a matter of formaldehyde, staining fluids, knives, slides, microscopes. And when it comes to enacting atherosclerosis as a limited walking distance in the out-

System or Episteme

Social theory used to ask this question: how is society ordered? How does it hang together, form a whole? The performances that Goffman studied hung together. They unfolded patterns that preceded them. "The pre-established pattern of action which is unfolded during a performance and which may be presented or played through on other occasions may be called a 'part' or 'routine' " (Goffman [1959] 1971, 27). These parts and routines added up into what Parsons used to call *roles*. There was a variety of roles, each coherent in its own terms. Together they guaranteed the coherence of the social system. The *social system:* the very term incorporates a specific answer to the question of how society avoids disintegration: society hangs together as in a system. In this respect it is just like the body. Or, better, just like the body was supposed to be in Parsons's day. When Parsons tries to explain what a *system* is, he inserts footnotes to the physiologists of the late forties and early fifties who took part in inventing cybernetics to account for the way in which the body hangs together.

But is a society like a body? In the same epoch, the idea was severely attacked. Can-

patient clinic, this includes the notes written down in a file: "reported pain-free walking distance: 150 meters." But it also encompasses the way the doctor looks (or doesn't look) in the patients' eyes during the interview and the patients' attempt to assess the distance they walk from their house to the park. To be is not only to be represented, to be known, but also to be enacted in whichever imaginable other way.

The word "is" used here is a localized term. Ontology in medical practice is bound to a specific site and situation. In a single medical building there *are* many different atheroscleroses. And yet the building isn't divided into wings with doors that never get opened. The different forms of knowledge aren't divided into paradigms that are closed off from one another. It is one of the great miracles of hospital life: there are different atheroscleroses in the hospital but despite the differences between them they are connected. Atherosclerosis enacted is more than one — but less than many. *The body multiple* is not fragmented. Even if it is multiple, it also hangs together. The question to be asked, then, is how this is achieved. How are the different atheroscleroses enacted in the hospital related? How do they add up, fuse, come together? In this chapter I will address the question of how the body multiple hangs together and present various *forms of coordination.*

One Reality Wins

Objects have local identities. But the various sites in the hospital where I went to study atherosclerosis are not entirely separate: the lower leg that is dissected in the pathology department is brought there by a special messenger who walked up all the way from the operation theater, also carrying a small paper on which the "clinical condition" of the patient before the operation is noted down. After

guilhem was among those who articulated a way of framing the difference. The norms that mark the order of an organism, he wrote, are given. But a society has to find regulatory norms and set them, actively. As Canguilhem put it: "In any case the fact that one of the tasks of the entire social organization consists in its informing itself as to its possible purposes . . . seems to show clearly that, strictly speaking, it has no intrinsic finality. In the case of society, regulation is a need in search of its organ and its norms of exercise" ([1966]

1991, 252). But though modern society *is* not a body, Canguilhem held that it *mimics* bodies. Bodies maintain their integrity by keeping up norms that mark the difference between order and chaos, life and death. These norms vary; in an organism that is pathological they are set at a different level than in one that is healthy. But if no norms are maintained at all, the organism becomes disorganized. It still obeys the laws of physics and chemistry, but becomes a biological chaos. It dies.

The relation between society and body

the pathology examination the results go back to the treating surgeon on yet another form. All the paperwork about any single patient comes together in a file. Summaries turn a specific patient's atherosclerosis into a single object. And so do letters. Here's one sent to a general practitioner about a patient who has been diagnosed, admitted to the hospital, operated, and discharged again.

> *Your patient D. Zestra*
> *date of birth: 13–04–1921*
> *address: Street 30, Smalltown*
> *hospital number: 2.892.130*
> *Dear colleague,*
> *Above mentioned patient was admitted in the department of Vascular Surgery, Surgical Unit, C 4-east, in academic hospital Z.*
> Admission facts
> *Date of admission: 01–08–1992*
> *Date of discharge: 08–08–1992*
> *Reason for admission: therapeutic intervention*
> *Diagnosis upon admission: stenosis in common femoral artery left with intermittent claudication*
> *Readmission: not planned*
> Anamnesis: *aggravating intermittent claudication with walking distance of 250 meters during which pain grew in left calf. There was no rest pain. The cardiac history was blank except for a hypertension.*
> History:
> *1981: bifurcation prosthesis and amputation fifth toe left*
> *1988: femoropopliteal autolog bypass of saphena magna vein left and right*
> *1992: amaurosis fugax with infarct left frontoparietal hyperlipidemia*
> Physical examination: *in the left leg the femoral artery was palpable. Distal of this*

is one of mimicry, but organizing a society by means of norms is not the only possible way of doing so: the normative mode of ordering is a historical invention, according to Canguilhem. Those who invented it didn't call the norms their invention, but claimed that they had found them as positive facts, in society. "Between 1759, when the word 'normal' appeared, and 1834 when the word 'normalized' appeared, a normative class had won the power to identify—a beautiful example of ideological illusion—the function of social norms, whose content it determined, with the use that that class made of them" (246). The idea was elaborated in Michel Foucault's work. "The Normal is established as a principle of coercion in teaching with the introduction of a standardized education and the establishment of the *écoles normales* (teacher training colleges); it is established in the effort to organize a national medical

point there were no pulsations. In the right leg all pulsations were present those of the foot included. The capillary refill left was slower than right.

Supplementary diagnostic examinations: *ankle/arm index left 0.6 and right 1*

Duplex: stenosis of more than 50 percent in the common femoral artery left

Operation: *(02–08–92) endarterectomy in common femoral artery left*

Further course: *no postoperative complications. Patient could be mobilized fast. The ankle/arm index was 1 right and left 0.9. Discharged in a good overall condition.*

Summary:

Main diagnosis: *stenosis in left common femoral artery*

Side diagnosis: *none*

Therapy: *endarterectomy in common femoral artery*

Complications: *none*

Discharge to: *home*

Control afterward: *outpatient clinic hospital Z*

With collegial respect,

Dr. T. F. J. Xanders, surgeon

A. J. Yielstra, surgery resident

What kind of disease did this patient suffer from? The letter mentions several diagnostic techniques that each gave an answer to this question: the anamnesis, a physical examination, pressure measurements, and duplex Doppler scan. They jointly back up a single diagnosis. This patient, they say, has a stenosis in the common femoral artery of his left leg. How is this remarkable alignment of such different diagnostic findings into a single diagnosis practically achieved?

Let's move back in the patient's itinerary, to a moment when the writing of

profession and a hospital system capable of operating general norms of health; it is established in the standardization of industrial processes and products. . . . Like surveillance, and with it, normalization becomes one of the great instruments of power at the end of the classical era" (Foucault 1979, 184).

Parsons had a theory about the contribution of doctors to the maintenance of the social system. It said that it is part of the sick role that the sick must seek medical assistance. Doctors subsequently either sanction their patient's illness be-havior or send them back to work again. It is in this way that physicians exert *social control*. They protect the social system from individuals who might want to enjoy the luxury of relief from their social obligations under the pretext of being unable to fulfill them. The Foucauldian concept of *normalization* also indicates that health care is important for the maintenance of social order. But Foucault's doctors do not control. They neither oblige people to stay in bed and get better nor to get up and go to work again. Instead, they set the standards of normality. They articulate what it

discharge letters is still far ahead. The decision about how to treat and whether or not to admit the patient still has to be made. There we are. In the outpatient clinic again. A vascular surgeon is seeing a new patient. The surgeon has written down this patient's walking distance and the results of his physical examination in the file on his desk. Both look serious. The *clinical diagnosis* is positive (positive for: disease present—instead of negative for: no disease found). The patient reported pain on walking, and the surgeon felt bad pulsations in several arteries. In the routine course of practice in hospital Z another diagnostic technique is now brought into play. The surgeon makes a phone call to check if the vascular laboratory is available, writes a note requesting the technician to check arm and ankle pressures of both legs, hands this note to the patient, and says: "Please, come back here afterward." If we accompany this patient we come across another mode of diagnosing, and delineating, vascular disease. Pressure measurement.

The technician measures the blood pressure in Mr. Manders's arm. She inflates a cuff around it. While she slowly allows the air to escape again she uses a stethoscope to listen to the artery in the elbow. An inflated cuff stops the blood from flowing. When some air has escaped, the sound of turbulent flow becomes audible. This is the moment the blood is able to push past the cuff when it is at its peak pressure, the systolic pressure. More air escapes and then the sound disappears again. It disappears at the point where the blood starts to flow undisturbed, able to resist cuff pressure all through the heart cycle. This second point is the blood's diastolic pressure. The technician writes both the higher and the lower number on a piece of paper.

She fits a larger cuff around Mr. Manders's ankle. In the ankle the stethoscope cannot be used. Instead, a small Doppler probe has to do the job. It sends out ultrasound and receives

is to be normal and to behave in a normal way. They may also actively intervene so as to bring about normal states. But unlike judges, doctors do not punish those who do not live up to their norms. Normality is not a law. Instead, those who do not manage to meet the standards of normality, the *abnormal*, are marginalized to the fringes of society. They come to find themselves in places where most do not want to be, places from which they will try to escape. Thus "normality" is something people come to positively desire, from the inside, instead of something that, like a rule, is imposed on them from the outside.

In framing his social theory, Foucault was not arguing with Parsons and other systems theorists. Instead he tried to stress—like Parsons, but without relating to him—that medicine is vital to society. It is a social power of a quite specific kind. "The power of the Norm appears through the disciplines. Is this the new law of modern society? Let us say rather that, since the eighteenth century, it has joined other powers—the Law, the Word (*parole*) and

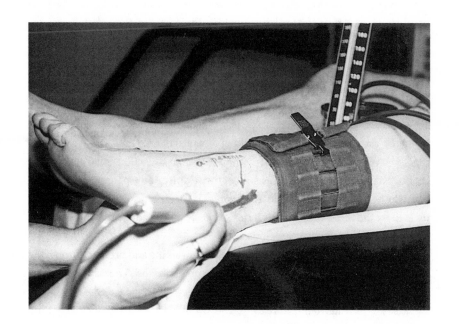

the reflections back again. If ultrasound is reflected by an object that is moving away, its reflections have a longer wavelength than the ultrasound emitted—and vice versa. This is the Doppler effect. The Doppler apparatus (to which the probe is attached) makes the difference between emitted and received waves audible. The technician moves the probe around until she's found the artery. We can all hear when she does, for at that moment the flow of blood reflects the ultrasound. "Pshew, pshew, pshew," we hear. When the cuff is inflated, this sound disappears. To come back again as soon as the systolic blood pressure is able to resist the cuff pressure. Movement. Flow. Pshew, pshew.

When blood pressure in an ankle is lower than in the arms, pressure is lost along the way. Like the pain-free walking distance of the clinic and the thickening of the vessel wall in the department of pathology, the *pressure loss* established in the vascular laboratory is a measure of the severity of the patient's atherosclerosis. In hospital Z, pressure loss is expressed as an index: the ankle pressure divided by the arm pressure. An index of 0.9 is used as the cutoff point: lower numbers are classified as pathological.

There is a story that explains how pain when walking and pressure drop hang together inside the body. The thick intima comes in, too. This is how it goes. When a thick intima encroaches the vessel lumen, resistance to the blood flow increases. This leads to pressure drop. The low blood pressure in the lower limb isn't high enough to supply the tissues with much blood. When the muscles are exercised the oxygen supply falls short. The muscles therefore burn their sugars without oxygen and produce lactic acid. This is painful. A convincing story. But does it hold? Well, it does as long as the various atheroscleroses enacted of a single patient all have more or less the same degree of severity. No, let's be more precise. We cannot know anything about the cross section of the arteries

the Text, Tradition—imposing new delimitations upon them" (184). In establishing the power of the norm, medicine is a crucial discipline, because medical knowledge mediates between the order of the body and the order of society. It is within medical knowledge that the normal and the deviant person are differentiated. It is within medical knowledge such as it has taken shape since the early nineteenth century that "disease" is no longer thematized as a species inhabiting an organism, but as a deviant state of that organism.

Since that time medicine has started to set the standards that modern people want to live up to. Thus, it is medicine that allows society to mimic organisms. And its own knowledge hangs together, too. It forms an epistème: a logically coherent *body of knowledge*.

This body of knowledge doesn't emerge out of isolated scientific activities and then invade society. New knowledge is not a product of clever minds. It emerges when scientific work is done in new sociomaterial settings. Foucault attributes inno-

of Mr. Manders. But his complaints and pressure drop are both investigated. When the results of these two diagnostic techniques coincide they jointly enact a common object. Mr. Manders's atherosclerosis.

When the technician has finished measuring, she takes a form. Mr. Manders is asked for his plastic hospital identity card and with a device that rolls over it so that the letters leave marks, Mr. Manders's identity is coded onto the form. Then the technician fills the appropriate boxes on the form with her findings. She tries to avoid mistakes. So she seeks confirmation from the patient: "It's in your left leg, isn't it, Mr. Manders?" Mr. Manders nods and says that, yes, it's in his left leg. He adds with a smile: "So, I could be a doctor of the blood vessels myself then. For I could feel it, that it's in my left leg." The technician is not impressed. "Of course you can feel it. I measure what you feel all right."

To Mr. Manders, the cuffs, the stethoscope, and the Doppler with its strange sounds that shift in tune with every heartbeat are pretty impressive technologies. He observes the work of the technician as attentively as I do. If doctors need the outcomes of all this work and equipment in order to know about his disease, Mr. Manders can be proud of himself. Or so he jokes. He needs no equipment at all to have access to his leg arteries. He can feel them.

The technician, however, sees nothing special in Mr. Manders's ability to feel what she measures. Complaints simply correlate with a drop in pressure because they are both signs of a single disease, hidden deep inside the body: he feels it, she measures it. Inside the body the one causes the other. Thus, their correlation is self-evident. Or is it? Complaints and pressure drop often coincide, but not always. Here's a second scene.

vative force to the novel organization of the French health care system at the beginning of the nineteenth century. This generated the birth of the clinic. It was with the specific hospital organization that emerged at that time that it became possible and reasonable to open up corpses in order to find disease inside them. Speaking about the new hospital organization, Foucault remarks: "It so happened that it was on the basis of this tertiary spatialization that the whole of medical experience was overturned and defined for its most concrete perceptions, new dimensions, and a new foundation" (1973, 16). Medical knowledge,

medical perception itself, is as social in its origins as in its effects. And it is material as well: a *discourse* that structures buildings, instruments, gestures. That differentiates between normal and pathological organisms and thus mediates between the coherence of the body and the order of society.

Associations and Multiplication

The idea that medicine is not just a personal affair between a doctor and a patient has never left the literature since. It has become commonplace, something we all know, a truism: that medicine is as social

Sometimes the clinic wins. The measurements of the laboratory are discarded. And this is the way to discard them: to *unbracket* the practicalities of measurement. To stop hiding, but include the activities of gathering knowledge about the body in one's story about it. To show what may go wrong there, for instance, by telling that it is among the specificities of a successful pressure measurement that the arteries are compressed when the cuff is inflated. If the patient's arteries are too calcified to allow for proper compression, pressure measurement loses its value. This can only be pointed out if one doesn't get mesmerized by the numbers that pour out of machines, but is prepared to take a step back in order to consider how such numbers are created.

Thus, though an object that hangs together inside the body tends to be established by *bracketing* the practicalities of measurement, sometimes this no longer works. Incoherence, however, can then often be kept at bay by *unbracketing* those same practicalities again. Practicalities that were diligently hidden are again attended to. If thus the gaps can be explained, the singularity of the body and its diseases is maintained. Don't trust tests, doctors therefore teach their students, they can fool you. Learn what they do. Get acquainted with their technicalities and know when to trust and when to discard them. This goes for all tests. Any single test outcome can be discarded. Explained away.

Two surgery residents are early for the weekly meeting where difficult vascular cases are discussed. One of them calls to the other and points at a small piece of paper. "Here, look at this. Have you seen the pressure measurements of Mr. Iljaz? It's unbelievable. I can't believe it. If you look at these numbers he can hardly have any blood in his feet at all. And he came to the outpatient clinic all alone, on his motorbike. Said he had some pain. I can't believe it. Some pain. On these figures alone I'd say here's someone who can't walk at all. Who's screaming."

on those who were not. Doctors with private practices, for instance, had nothing to gain from following Pasteur. So they didn't. They preferred to maintain confidentiality in their relations with patients and refused to tell outsiders whom to vaccinate. Even when the first serums were produced doctors did not prescribe them, for in order to do so they would have had to hand their patients over to other professionals. Private doctors only started to "believe" in serums once the Pasteur laboratory put these on the market, and the doctors were free to use them in their own surgeries when they considered it appropriate.

So "science" doesn't have the power to impose itself. If it spreads this is because there are actors outside the laboratory who associate themselves with it. And they may pick through what is on offer and take bits and pieces. They do not get overwhelmed by a massive structure or a coherent epistème. Latour talks about *chains of associations* instead. Chains

Here the clinical diagnosis is doubted on the basis of the laboratory numbers. Sure, in the clinic Mr. Iljaz was diagnosed as someone with a probable arterial disease in his legs, that is why he was sent to the vascular laboratory in the first place. But the clinical picture wasn't dramatic. Mr. Iljaz still walked and drove around on his motorbike. He reported pain but not agony. His pressure measurements, however, show a very severe degree of atherosclerosis.

The incoherence is big enough to warrant an explanation. What can it be?

The others enter the meeting room one by one. A senior internist has joined the residents. "Yeah, that really is something," he nods, "but we've seen cases like that before. Probably these people have only become worse very gradually. What happens is that their muscle metabolism alters. As long as people have time for it, the adaptation may go a very long way. And then. What's this patient, let's see. Does he have diabetes? For that is also something to bear in mind. If he's developed a neuropathy, he may no longer feel any pain at all. It happens that you can stick a needle in peoples' feet without them even blinking."

Crucial to enacting a clinical diagnosis is the patient's capacity to feel pain. The patient may feel no pain if his movements have slowed down and his muscles have adapted to a low level of oxygen. And a patient doesn't feel pain either if his nervous system is in a bad state due to long-standing diabetes. A limited capacity to feel pain may explain the discrepancy between clinical findings and pressure measurement. But there are more possible explanations for the gap. Clinical diagnosis, after all, doesn't simply depend on the patient's body, but also on the clinical interview. It is quite difficult to do this well.

After the meeting a student asks the resident who was on duty in the outpatient clinic: "Does this Mr. Iljaz speak proper Dutch, or did you have an interpreter?" The resident

that form networks. These may be long or short, strong or weak. Their coherence is a material and a practical matter, not a question of logic. Strength depends on what sustains the associations. It is defined by the activities required to disrupt them and bring about fragmentation. "The consistency of an alliance is revealed by the number of actors that must be brought together to separate it" (206).

Latour dissolves the power of logical coherence by arguing that in as far as the world hangs together this is a matter of practical associations. How far these associations reach isn't given with the birth of a new configuration. Unlike epistèmes, networks are open. The elements within a network may link up with other elements, outside the network. But such external links are not different from internal links. They're all associations. Each new and successful association makes a network larger. But however great the difference between the coherence in a network and *logical* coherence, to talk of "associations" does have a homogenizing effect. Either an association

sighs. "Yeah, come to think of it, he may have underreported his complaints. His Dutch was poor. And I didn't have a lot of time either. You're right, with someone to translate we might do better in cases like this. Well, I'll try to talk to him properly as soon as we've got him admitted. Ask some family member to help. Or, indeed, the interpreter."

Lab outcomes and the results of a clinical diagnosis are supposed to line up. But sometimes they don't. Then it requires some coordination work to still align them. Ask a few questions. Were the arteries too hard to be compressed by a cuff? Did the lack of blood come about so slowly that the muscles adapted? Were the patient's nerves in a bad state? Did the language that doctor and patient used during the interview suit both—or was only one of them fluent in it? In the specificities of the practicalities of enacting a disease, an explanation may be found for the inconsistency of two diagnoses. One of them wins. The other is discarded. Thus a single patient ends up with a single atherosclerosis.

A Composite Picture

Do patients always have a single atherosclerosis? Do the names of individuals always come with coherent bodies? No. It is more complicated. When different tests give different outcomes, it is not obligatory to abandon one. It is also possible to understand the objects of two different techniques as indeed being different objects. In such a scheme both *pain when walking* and *pressure drop* are troubles that may plague a patient. Troubles that have a relation, but not necessarily one that is linear. Troubles in their own right.

I found an intriguing example of this in an article that reports on the effects of treatment of arterial disease. The effects of two treatments are compared. The

is made or it isn't. An element is either inside or outside a network. Coordination is established or not. There are no distinctive *forms* of coordination.

The second way of abandoning Foucault differs from the first in precisely this respect. It multiplies. Instead of describing a single coherent discourse, or tracing a single large network of contingent associations, it distinguishes many . . . *Many what?* There are different answers to this question in the literature. Different ways of multiplying have established themselves, side by side. And there is yet another

complication: even if some of those who multiply come *after Foucault* in the sense that they multiply what one might still call discourses, others draw on quite different traditions. Intellectual history isn't like a single tree with endlessly subdividing branches. Instead, there are overlaps, resonances, shared topics, and crossovers between traditions that are quite alien to each other in other respects.

So how to relate to these widely spread and equally relevant literatures? I'll make a list. A list of multipliers.

first treatment is percutaneous transluminal angioplasty (PTA), which works by inserting a balloon in a vessel and inflating it to widen the lumen. The second treatment is exercise. We'll come to talk of both of these in the next chapters. What only counts for now is that in the study, each of these treatments appeared effective. But they had different effects.

"In a prospective randomized trial, Creasy et al. (1) compared the results of percutaneous transluminal angioplasty (PTA) and exercise in the treatment of intermittent claudication. In patients treated by PTA a significant rise in ABPI [ankle/brachial pressure index, a way of expressing the outcomes of pressure measurement, in this book referred to as the ankle/arm index] was seen without increase in maximum walking distance, whereas those who received exercise training showed a significant increase in maximum walking distance without an increase in ABPI." (From van der Heijden FH, Eikelboom BC, Banga JD, Mali WP: The management of superficial femoral artery occlusive disease. British Journal of Surgery 1993; 80:959–996; reference (1) cited is Creasey TS, McMillan PJ, Fletcher EWL, Collin J, Morris PJ: Is percutaneous transluminal angioplasty better than exercise for claudication? Preliminary results from a prospective randomised trial. European Journal for Vascular Surgery 1990; 4:135–140).

One treatment, PTA, improved the ankle pressure. The other treatment, exercise, improved the patient's walking distance. In the study quoted here, both indicators of the degree of the patient's vascular disease were measured. Contrary to the expectations they did not run parallel. What to do? Discard either one of them? There is another option. It is to say that if they do not run parallel they may be objects in their own right. Different objects.

Sometimes this is done. The outcomes of two diagnostic techniques are drawn out of their signifying role. Instead of signs of a single atherosclerosis underneath, they are accepted to be what they are on the surface. Pressures — or complaints. If they differ, neither needs to be abandoned. Their difference

1. There are those who talk of *social worlds*. Social worlds are groups of people who share perceptions and ways of talking about them. They have similar interpretations and attribute similar meanings to the events they encounter. Surgeons and social workers may belong to different social worlds. Or lay people and professionals. Or scientists and clinicians (Strauss 1978).

2. Others distinguish between *versions* of the world. Like social worlds, versions are perspectival in character, they are ways of interpreting, but they do not neatly overlap with groups of people. A single person may be both a physicist and a musician and thus be engaged alternately in the ways of worldmaking of physics and music (Goodman 1978).

3. Individuals do not coincide with the next multiplier either: the *frame*. People may draw on various frames, depending on the specificities of a situation. In so-

implies no incoherence, for the two measurement techniques do not assess the same disease. They each have their own object. In this sense a single patient may now be diagnosed as having two "atheroscleroses," pain on walking and pressure drop. These two objects do not necessarily coincide.

If the outcomes of diagnostic techniques are taken to stand for different objects, however, these may be aligned again to form a single one. The form of coordination that comes into play here is that of adding things together. Don't bother about whether they're *really* similar or different. Don't try to explain how they hang together inside the body. Forget about the body. Just add up your findings. With pressure drop and pain, the "criteria for success according to Rutherford" make exactly this move. In the Rutherford calculation, indicators of success are not played out against each other, but added up. If one is positive and the other negative, neither has to be discarded. They can even be substituted for one another.

In the literature, the "criteria for success according to Rutherford" are used over and over again. Not only by Rutherford himself, but by many others as well. This allows comparison between different studies that evaluate treatment outcomes. In the "criteria for success according to Rutherford" improvement is defined in a composite way. It is a combination of clinical symptoms and ankle/arm index. Various categories of improvement are differentiated. For example, the best score is +3, markedly improved. This is scored when (a) symptoms have either disappeared or markedly improved, while (b) the ankle/arm index is increased to more than 0.9. The most striking addition, however, is improvement category +1, minimally improved. This is scored when (a) the ankle/arm index is increased more than 0.1, while (b) the symptoms have not made a jump from one symptom category to another, or vice versa (F. van der Heijden: Semiclosed endarterectomy of the superficial femoral artery. Thesis, Utrecht, 1994).

cial medicine, for instance, two frames can be distinguished. There is a clinical frame, held together by the aim of "helping people," and an administrative frame held together by the aim of distinguishing between the "objectively sick" and other people. These imply two ways of interpreting but also two ways of acting: asking questions, filling forms, doing a physical examination (Dodier 1994).

4. And then there are *modes of ordering*. Modes of ordering do not primarily order meanings (like "versions") or actions (like "frames"). They have neither a thinker/feeler nor an actor at their center: individuals are ordered along with them. Modes of ordering pervade organizations, or habits, or buildings, or techniques, or gestures. They may order anything: what it is they order is part of what turns them into one "mode" or another (Law 1994).

Out of all these multipliers "modes of ordering" most resemble Foucault's "dis-

The patient's situation is defined as improved as soon as one indicator has improved even if the second indicator has not changed. An increase in ankle/arm index without improvement of the clinical symptoms: fine. Or vice versa: less severe clinical symptoms but no change in the ankle/arm index: fine again. No indicator is discarded, no gap between them warrants explanation. Treatment has been for the better if *either* pressures *or* walking distance are improved.

Addition is a powerful way of creating singularity. This becomes clear at the moment surgeons do not ask "what is the matter?" but rather "what to do?" Because when the vascular surgeons of hospital Z try to decide "what to do," they are not only interested in complaints and the results of pressure measurement. They draw in a lot more elements.

The man sits on a chair facing the surgeon. He's 84. He lives in a home. He looks worn out. Tired. "Listen, Mr. Winter," the doctor says, "there is indeed something wrong with your arteries. That's what we've found out in this examination you've had, the pressure measurement. I've got the numbers here. They're not very, very bad, but they're bad. Maybe we are able to do something about it. I can't promise. But if we can, it's either with a small balloon, or with an operation. We need more information to know what's possible. But we are only going to submit you to more examinations if you would actually want treatment. If you don't, well, that's fine. You're not in danger or anything. The treatment, if it is possible that is, the treatment would just be to allow you to walk better. So maybe you could think about that, about whether you would want treatment."

Mr. Winter's pain-free walking distance is some 120 meters. The ankle/arm index of his right leg is 0.7. These findings are added up to the diagnosis "impaired blood flow in right leg." They are enough not to send the patient back

courses." Look at the list I've just presented: it follows the decentering of the subject. The subject shifts from a central sense maker, to a decentered sense maker, to an actor centralized by the analysis, to a being performed in various modes of ordering. But this doesn't imply that "modes of ordering" are simply "discourses" multiplied while everything else has stayed the same. John Law claims to have taken several steps while abandoning the Foucault he has digested. And he suggests that his readers do so with him. "My proposal is that we take the notion of dis-

course and cut it down to size. This means: first, we should treat it as a set of patterns that might be imputed to the networks of the social; second, we should look for discourses in the plural, not discourse in the singular; third, we should treat discourses as ordering attempts, not orders; fourth we should explore how they are performed, embodied and told in different materials; and fifth, we should consider the ways in which they interact, change, or indeed face extinction" (95).

So where can one go after the *discourse?* In the literature there are two great roads

immediately to the general practitioner, but instead to consider treatment. Is treatment an option? This depends on a series of further assessments. The first of these is social. Does the patient's bad right leg seriously hamper him in his daily life? Maybe he lives in a home, is taken care of, and hardly ever wants to go out anyway. Then invasive treatment is unlikely to improve his life. What does Mr. Winter think about it? Does he think an operation might improve his life enough to be worth the hospital admission, the suffering, the recovery time, and the risks of getting worse instead of better? Is he *motivated*?

The "atherosclerosis that requires invasive treatment" is a composite object. The social reality of living with atherosclerosis is included in this patchwork. *Social atherosclerosis* is added to the other versions of the disease. There is no expectation of a linear relation between walking distance or ankle/arm index and "disturbance of daily life" and/or "motivation." It is precisely because nobody expects there to be a linear relation between a patient's physical disease and what we might call his "social disease" that the latter deserves separate attention. Thus, the fact that different objects may be added together and thereby turned into one doesn't depend on the projected existence of a single object that was waiting in the body. Singularity can also be deliberately strived after. It can be produced. The *result* of addition is a single object. An atherosclerosis that should be treated invasively. Or not.

Coordination into singularity doesn't depend on the possibility to refer to a preexisting object. It is a task. This is what designing treatment entails. That the various realities of atherosclerosis are balanced, added up, subtracted. That, in one way or another, they are fused into a composite whole.

In the decision-making meeting, the test results of Mrs. Stienstra have been laid out. I was in the outpatient clinic when Mrs. Stienstra came in. A sociable woman, well into her

to follow. One is a product of doubts about the *force* by which a discourse hangs together as a whole. This doubt leads to the invention of *networks* that gradually come to hang together by means of small forces—forces that the analyst cannot presume to be there, but must be able to point out: *associations*. The other road is paved by doubts about the *extent* of the discourse that hangs together. This leads to the pluralization of a single order into different coexisting, no, not *orders* but, in proces-

sual terms, *modes of ordering* that interact, change, or face extinction.

You may read the present book as wrestling with some of the questions raised by these two ways of building on—but also moving away from—Foucault. A first question: it may be that, at least in each empirical study, it is possible to follow the associations made within a single network. But what if there are two or more networks? How then to articulate the difference between associations *within* and *between* net-

seventies. Eager to live, to go out. Outgoing. "Now, what's this?" a senior surgeon says to a younger one, pointing his finger at some test result. "You want to operate on this patient? And she has these vessels? Are you serious? It's not that bad, is it? Why don't you just wait and see and tell her to keep moving?" The younger one stays calm. "Well, yes. You may be right. But the problem is: this woman so much likes to go to places. Last year she still traveled and now she can't and it's what she lives for. So why not give it a try?"

Another reality is added to the results of various tests. Social atherosclerosis. It may be important, or not. It may have a lot of weight, or little. It has some striking similarities with what sociologists who investigate it with their own techniques call *illness*. What are the techniques for enacting social atherosclerosis in the consulting room? They are difficult to handle. Difficult for patients who have to tell appropriate stories, show well-balanced emotions, be articulate. And also difficult for the doctor who has to ask good questions, listen carefully, and even understand what isn't articulated. The social reality of living with disease may be so bad that some patients would rather die than undergo more treatment. They may say that in as many words. But it also happens that they say nothing at all.

Surgeon to resident: "Yeah, I've seen the results for that patient. What's he called again? Vandervoort. You're right: his pressures are bad. Yes. There can't be much of a lumen left. But somehow I don't believe we should do an intervention. I don't think he wants any of this. It's his children that want treatment. They do all the talking. To tell you the truth, I don't really know how to proceed. If it weren't for these children, I would already have stopped this whole circus. But if I do nothing, and he deteriorates, which he's likely to do, well, we can expect his children to get angry."

The "disease to be treated" is a composite object. The elements that compose it may stretch all the way from the numbers that come out of the vascu-

---

works and—more important still—might it be the case that different networks hang together in different ways, are there different *kinds* of association? And this is a second question: what turns one mode of ordering into a *mode* of ordering and what terms might we use for the way in which it differs from another? These two questions, then, inform my inquiries into forms of *co-ordination* between different enactments of atherosclerosis in hospital Z.

Paradigm

In 1962 Thomas Kuhn published *The Structure of Scientific Revolutions* (Kuhn 1962). A quote. "An investigator who hoped to learn something about what scientists took the atomic theory to be, asked a distinguished physicist and an eminent chemist whether a single atom of helium was or was not a molecule. Both answered without hesitation, but their answers were not the same. For the chemist the atom of helium was a

lar laboratory to the possible future anger of someone's disappointed children. Such different elements together make a patchwork. A patchwork singularity, the disease-to-be-treated of a specific patient. A composite reality that is also a judgment about what to do.

Translations

When clinical findings and pressure measurement suggest it might be worthwhile to engage in invasive treatment, the vascular surgeons of hospital Z ask their patients about their daily life and try to find out whether or not patients are motivated for invasive treatment. "Would you want invasive treatment?" they say. Only if the patient's social atherosclerosis is bad enough and if treatment is likely to improve the patient's situation will more facts be assembled. More facts. The design of treatment is not just a matter of (a) establishing the presence of vascular disease and (b) establishing the necessity of invasive treatment. There is a third necessary ingredient. In order to choose an appropriate invasive treatment the patient's vascular disease must (c) be *localized* and *quantified*.

In the next chapter I'll pay more attention to the various invasive treatments and their indication criteria. Now it is enough to know that in the design of treatment the *site* and the *size* of a patient's vascular disease are important. How do the surgeons of hospital Z find out about these? There are several possibilities. Here I'll focus on two of them, duplex Doppler and angiography, and on some of the similarities and differences between their objects. Angiography is the older of these diagnostic techniques. It is invasive.

They stand around the patient. There's three of them. They're clad in sterile green. They wear aprons to protect themselves against the X-rays and gloves to protect the patient against their microbes. The moment the needle finds the artery in the groin is tense. Yes. There it is.

molecule because it behaved like one with respect to the kinetic theory of gases. For the physicist, on the other hand, the helium atom was not a molecule because it displayed no molecular spectrum. Presumably both men were talking of the same particle, but they were viewing it through their own research training and practice" (50–51).

With this story Kuhn illustrates the nature of a *paradigm*. A physicist and a chemist live in different worlds and answer

simple but vital questions differently. It fits within the chemist's research training and practice to call a helium atom a molecule. But within the physicist's research training and practice it doesn't. They work within different paradigms. When it was coined, the term *paradigm* first helped Kuhn to move out of a *fragmented* world. Out of too radical a pluralism that separated the building blocks of science out into independent *sense data*. Many a philosopher of science in Kuhn's day took *sense data*

Blood spurts out. Then the puncture in the artery is blocked by the catheter. This is pushed in and pushed in. It moves down. On the monitor its movements can be followed. There it goes. Good. Now stop it. Yes. It's where it should be. A technician approaches and attaches the automatic dye injector. The patient lies pale and sedated on the examination table. One of the residents addresses the patient's head. "Could you please lie quietly, Mrs. Lensi? You will get a sudden warm feeling in your leg, that's all right, that's how it should be. But if it hurts, please call us and we'll be with you straight away, all right?" Leaving Mrs. Lensi alone on the examination table all the others—one radiologist, two residents, a visitor from Switzerland who wants to learn new methods, a technician, and me—retreat to the adjacent room. Here the buttons of the impressive X-ray machine can be pushed. Clack. The dye injector injects dye. And immediately a card with holes in it sets the X-ray machine taking pictures. Paff, paff, paff. One after the other.

An angiographic picture shows the lumina of the arteries downstream from the place where dye has been injected. Like bone mass, the dye used in an angiographic procedure is opaque to X-rays. It casts a shadow on the X-ray plate. Angiographic images thus show the lumina of the vascular tree below the point of injection in a two-dimensional, anatomical mode. The site of the stenosis can be pointed out with a finger and expressed in the anatomical language in which parts of arteries each have their own technical name. But the size of the disease is more difficult to assess. It is expressed in percentages lumen loss.

Decision-making meeting. The light box. A surgeon walks up to the angiogram under discussion. "How much did you make of this?" he asks the radiologists, his finger pointing toward a stenosis. "Seventy percent? Come on, that's not 70 percent. If you compare it with

to be devoid of meaning. They were literally data: given to the bare and naive senses. But nothing, or so Kuhn argued, is devoid of meaning. Data aren't isolated entities floating around in a homogeneous void. The senses only perceive what makes sense to them. And only that which fits with earlier perceptions and with theories about them may hope to make sense. The only exceptions to this are a few anomalies that linger in the margins until, one day, they fit into a new paradigm.

Thus, paradigm is a term that designates connectedness. The connectedness within physics or within chemistry. Or the connectedness within Aristotelianism. But it was precisely the connectedness inside these paradigms that made it possible to articulate the differences between them. And made it clear that the radically pluralist world where all sense data float independently, is, paradoxically, homogeneous. The sense data that a scientific theory was supposed to draw together come from a place that in being devoid of meaning, is, indeed, a void. There is no relatedness, and thus no difference in it. Pointing out the relatedness between *some* data but not others led to cleavages in this homogeneous *whole* of science. Physics

the earlier part there, if you take that bit as the normal part, up here, I'd say it's almost 90 percent, this lumen loss."

Despite the high interobserver variability (the official name for such disagreement) practicing vascular surgeons have little problem with the accuracy of angiography. They listen to the judgments of the radiologists and, moreover, interpret the images themselves. In the end they come to some conclusion. They do not need a reproducible fact: what they need is a decision. Angiography helps them to decide.

And yet a new diagnostic technique has made its way into the diagnosis of arterial disease. Angiography involves risks for the patient: some people are allergic to X-ray dyes, they may get very sick or, in rare cases, die. Others are left with a large blue bruise in their groin from the puncture. After classical angiography it is therefore necessary to monitor the patient—and this implies hospital admission. The younger technique, duplex, has none of these problems. It is noninvasive.

A small room. A patient, Mr. Fransen, lies on an examination table. Next to his head there is a large apparatus with lots of buttons, two monitors. Out of the apparatus come the cords of several probes. A technician moves one of these probes over Mr. Fransen's abdomen. His legs. From time to time the technician squeezes some gel between probe and skin, to conduct the ultrasound that the probe sends out and receives back again. There's little talk. The technician only looks at his right hand once in a while. Most of the time he is silently watching the screens. There are white shadows: echos of ultrasound reflected by tissue. Sometimes a vessel. He aims his probe at its interior and red and blue become visible: flowing blood. Flowing blood reflects ultrasound with a different wavelength than what has been emitted.

and chemistry do not link up with one another smoothly; there is a gap between them, as there is between the Aristotelian paradigm and that of Newton. They are incommensurable. There is no longer an atomic plurality of data that comes with a homogeneous science: the connectedness within paradigms comes with differences between them.

The differences between paradigms are unlike those between sense data. Incommensurability doesn't imply that the borders between paradigms have no crossing points. *Translations* may be possible. In some cases. Such translations require not only linguistic skills. The human senses involved have to be able to perceive different data as well. They must be able to make a *Gestalt switch*. And if the data depend on instruments, which in modern sciences they tend to do, then these instruments must likewise allow for translations. Sometimes they don't. That is not a matter of attributing meanings, but one of doing experiments. It is a practical matter. Ian Hacking puts it like this: "New and old theory are incommensurable in an entirely straightforward sense. They have

The colors on the screen represent the blood's velocity. If it is red it is flowing away from the probe; if it is blue it is approaching it. (But the apparatus allows the technician to adjust the colors.) The apparatus is also capable of making this information — velocity — audible. Pshew, pshew. Shifts in velocity over every heartbeat. And it can be represented as a graph. There, on one of the two small screens, look at it. The plot of an average graph of several heartbeats. Every so often the technician pushes a button and the images on the screens are saved and printed.

The ultrasound emitted and received back by the duplex apparatus has no known side effects. The gel is soluble in water. It makes things easier if the patient does not eat before an examination that involves the arteries in his abdomen, but after the hour, two hours that the examination takes, the patient can do whatever he wants again. Eat, go home on a motorbike or even a bicycle.

So the new technique is more "patient friendly" than the older one. But this is not enough to make it acceptable as a diagnostic device. Vascular surgeons do not only want a safe diagnostic technique, but also one that is reliable. Are the outcomes of duplex as good as those of angiography? Duplex and angiography present different data. An angiographic image shows the vessel lumen, and duplex tells about blood velocity. The objects of these two techniques are different. How then can duplex be *as good as* angiography in assessing atherosclerosis? How to compare the width of a vessel lumen with blood velocity?

In order to coordinate their outcomes, duplex and angiography were *made* comparable. This work was well under way when I began my fieldwork. One of my informants in hospital Z defended a thesis about it.

"The aim of this thesis was to study the ability of duplex scanning to accurately assess stenoses and occlusions of the aortoiliac and femoropopliteal arteries in patients with atherosclerotic disease, and set proper diagnostic criteria for the detection by duplex scanning

no common measure because the instruments providing the measurements for the one are inapt for the other. This is a scientific fact that has nothing to do with "meaning change" and other semantic notions that have been associated with incommensurability" (Hacking 1992, 56–57).

Science, or so Hacking states, is not unified "in part because phenomena are produced by fundamentally different techniques" (57). The plethora of techniques makes for a multiplication of reality. The

unification of the sciences is no longer viable, not even as a promise at the horizon. "We staunchly believed that science must in the end be unified, because it tries to tell the truth about the world, and there is surely only one world. (What a strange statement, as if we had tried counting worlds.)" (57). We staunchly believe*d*. Ian Hacking puts it in the past. "We" no longer believe that data are independent of the technology that makes them. Thus, since there are many techniques, there are many

of haemodynamically significant arterial lesions" (D. A. Legemate: *Duplex scanning of aortoiliac and femoropopliteal arteries*. Thesis, Utrecht, 1991, p. 95).

The ability of duplex to accurately assess stenoses and occlusions was established by comparing duplex outcomes with those of angiography.

"In this prospective study in 61 patients duplex scanning was compared to angiography in the assessment of atherosclerotic lesions of the aortoiliac and femoropopliteal arteries" (D. A. Legemate: *Duplex scanning of aortoiliac and femoropopliteal arteries*. Thesis, Utrecht, 1991, p. 60).

So how to correlate duplex outcomes with those of angiography? A duplex graph shows the changes in blood velocity over the beat of a heart. Various parameters may be derived from this. The flow profile, for instance, or the height of the graph's peak (the peak systolic velocity [PSV]). Or total flow: it might be possible to try that, to find out the vessel diameter with the echo and to measure the area beneath the curve, to then calculate the total blood flow.

But the favorite duplex parameter of my informants was the PSV ratio. This is the ratio of the peak systolic velocity inside a stenosis and the peak systolic velocity in a normal part of the same artery just before or just after the stenosis. The PSV ratio is a relative value, a matter of increase only; the absolute velocities are calculated away. It is this parameter that, in the quoted study, was correlated with the outcomes of angiography. Once a parameter was picked, the question could be asked whether its values were the same or different as the outcome of angiography. But how to compare PSV ratios with lumen loss? What might be their common measure? In the study quoted this problem was solved by dividing the angiograms of the sixty-one patients involved into three categories: lesions lower then 50 percent lumen loss; lesions between 50 and 99 percent

worlds as well, even if it makes no sense to try to count them. In theory, and with examples mostly drawn from physics, Hacking has outlined the technique-dependent multiplicity of *objects* that forms the topic of the present book already quite a while ago. But luckily there is something left to develop, for Hacking hasn't talked about how to separate out "science" when we no longer believe in its unity, nor about how different knowledges manage their coexistence.

Is it wise to talk about the disunity of science with this term, *paradigm*, that has come to be such a popular one for doing so? In order to answer that question, I'll take you to a very different part of the literature, one that, in the sixties when Kuhn wrote his well-placed intervention, was quite far removed from the discussions about science in which he intervened. However far away the literatures about science and society were those days, "paradigm" resonates with a specific so-

lumen loss; and occlusions. Then the tinkering started. Would it be possible to find cutoff points for the PSV ratio that would allow duplex to divide the same group of patients in more or less the same way in these same three categories? The answer was yes. It turned out that a PSV ratio of 2.5 proved a good cutoff point for differentiating between lesions of more and less than 50 percent lumen loss.

"Although in some studies a PSV ratio of 2 has been used to differentiate between stenoses of less or more than 50%, analysis of our results showed that this value would have given a markedly lower positive predictive value (64%) than a PSV ratio of 2.5 (ppv 82%)" (D. A. Legemate: Duplex scanning of aortoiliac and femoropopliteal arteries. Thesis, Utrecht, 1991, p. 96).

A PSV ratio of 2.5 or more has a correlation (a positive predictive value, to be accurate) of 82 percent with a loss of more than 50 percent of the vessel diameter as assessed on the basis of angiography. This turns a PSV ratio of 2.5 or more into a good parameter. One that is better than a PSV ratio of 2 because it correlates with angiographic findings better. Duplex findings are given meaning by setting up ways to translate them—a PSV ratio larger than 2.5 for instance—into angiographic findings—a stenosis between 50 and 99 percent. The possibilities for quantification are thus established—duplex can quantify arterial disease—and simplified—the quantification is no longer a scale with small gradients of difference, but a matter of classification into three groups. And this is how the objects of angiography and duplex are coordinated into a single common one: the severity of some patient's stenosis.

In the room in hospital Z where technicians write down their duplex find-

cial scientific way of articulating connectedness. Paradigm resonates with *culture*. However much Kuhn claimed he was after something different, the two terms have a similar way of drawing some things together into a coherent whole and thereby differentiating them from others. They turn what might have seemed to be isolated fragments together into wholes *and* postulate that instead of a single homogeneous universe we inhabit different worlds. "The concept of culture used by anthropologists was, of course, invented by European theorists to account for the collective articu-

lations of human diversity. Rejecting both evolutionism and the overly broad entities of race and civilization, the idea of culture posited the existence of local, functionally integrated units. For all its supposed relativism, though, the concept's model of totality, basically organic in structure, was not different from the nineteenth-century concepts it replaced. Only its plurality was new" (Clifford 1988, 273).

The term *culture* indicates plurality. But within each culture, again and again, there is—there was—a relatedness that resembles that of the organism. That is why

ings, the shortest possible summary of the thesis quoted here is printed on a page: the various PSV ratios and the "lumen loss" with which they correlate.

PSV ratio smaller than 2.5: a stenosis smaller than 50 percent. PSV ratio equal to or larger than 2.5: a stenosis larger than 50 percent. No sign: occlusion.

This translation rule *submits* duplex to angiography. It does not submit a given duplex graph to an angiographic image once they have both been made available and evaluated. Instead it submits the very way duplex graphs are read. Some of the proponents of duplex are critical of this submission.

Physiologist who has done duplex research: "They wonder what information they can get out of duplex by comparing it with angiography. Doing so, they accept angiography as the gold standard. But there are a lot more problems with angiography than with duplex. Angiography shows only two dimensions, it shows the vessel diameter, but not the entire surface of a lumen. And it expresses the severity of a stenosis as an index: a percentage of loss. But in arteries that were small to begin with a 50 percent loss is far worse than in larger arteries. Then there's the interobserver variability. Sure, duplex is technician dependent. If the technician misses a stenosis it can never be retrieved. But once you have a good technician, duplex outcomes are far easier to replicate. In angiography different observers never get to agree."

However fierce such criticism may be, in hospital Z duplex outcomes are nevertheless translated into percentages of lumen loss, which means that duplex is made to speak about the same object as angiography. Both technologies can be used to localize and quantify this object: a patient's stenosis. It is in this way that duplex gradually became an understandable, acceptable technique. But translations are never smooth. The study quoted talked about an overlap of 82 percent

---

Clifford, in the late eighties, tried to get away from it. And this indicates that by that time an era was ending that had begun, or so Marilyn Strathern tells us, early in the twentieth century with, or after, Morgan. "Morgan belonged to an era that had just finished debating whether humankind had one of many origins; Clifford speaks for a world that has ceased to see either unity or plurality in an unambiguous way. What lie between those are years of modernist scholarship with their vision of a plurality of cultures and societies whose compari-

son rested on the unifying effect of this or that governing perspective. Each perspective simultaneously pluralized the subject matter of anthropological study and held out the promise of a holistic understanding that would show elements fitted together and parts completed" (Strathern 1992b, 111).

Strathern tries to develop what it might mean to see neither unity nor plurality in an unambiguous way. Doing so, she criticizes the image of *fragmentation*, since fragments suggest regret about a whole

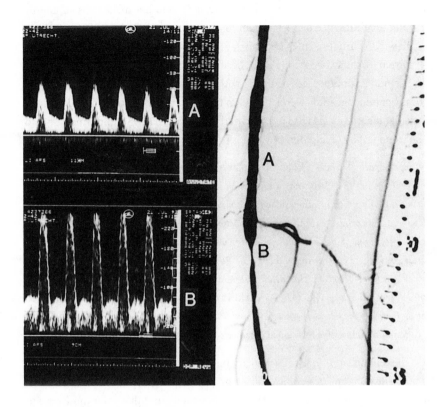

between the patient categories diagnosed by duplex and angiography. A difference of 18 percent is accepted as a part of the bargain of gaining the availability of the safe duplex as a diagnostic tool. Correlation studies make tests similar by taming the difference between them to a "reasonably small percentage."

In written articles, rules for translating duplex parameters into percentages of lumen loss may be established. These rules can then be used in the hospital. In hospital Z, photocopies of a page with translation rules that came out of the locally conducted study were at hand in the vascular laboratory. In addition, the vascular laboratory also has a visual method for translating duplex findings into the iconography of angiography.

While the patient dresses I follow the duplex technician to another room. Here light boxes make it possible to look at images printed on transparent plastic. The technician takes his prints out of a machine. They show the duplex graphs and the white echoes with red and blue that were printed each time he pressed his button. He looks at them carefully. Then he takes a form out of a stack. It allows him to write down psv ratios for various parts of the artery. And there's a drawing in the middle. It shows the aorta and the larger leg arteries in schematic form. The technician draws a stenosis in this image: he enlarges the vessel wall of the left femoral artery with a blue pen. Marks the picture until half of its lumen has gone, at more or less the height (he notes the amount of centimeters above the knee) where he's just found an impressive increase in blood velocity.

The technician translates a velocity increase into a loss of vessel lumen with a pen. The outcome of a duplex test—a graph, hard to grasp—a psv ratio, a number with as yet no meaning—is translated into a picture that is far more easy to read for those who are used to angiography. It is a pencil drawing in which the color blue represents what in an angiographic image would be a white shadow. If ever there were one, this is a translation.

that has exploded. She also criticizes the separating out of *elements* that may combine in whatever way they please, since this image evokes isolated genes that inherit independently, leaving the offspring with bits and pieces of both lines of ancestors. She wants us to get away from traditional *scales* in which the local is part of something larger, an encompassing globality. But how to get away from the idea that there are cultural packages, coherent inside and different from what is elsewhere?

One of the counter images that Strathern mobilizes is that of *partial connections*. It alludes to what, not in itself but through the act of comparison, appears to be both similar and different. Not like a single large cloth that is cut into smaller pieces after which the lost unity is simply a form to be sought. Not a functional unit nor an antagonistic opposition. But inside and outside. Strathern gives the example of the scholar who is simultaneously a feminist and an anthropologist. Being one shapes

ACADEMISCH ZIEKENHUIS
Afd. Radiodiagnostiek / Vaatdiagnostiek
Toestelnr. 6696 of 6763 Huispostnr. E 01.341

Datum: 18/6/97. Huispostnr.: _____
Onderzoek: Rgcr Ceph.
Tijdsduur: 30 . ~. Afdeling: C620.
Aanvrager: Claud. int.
Klinische gegevens: Buza

? 0

5 - 0 - 39

ECHO DOPPLER ONDERZOEK
BEKKEN/BEEN ARTERIËN

	PSV	PSV^R	RATIO	% STENOSE	R
AO					
	80				
AIC	60				
	80				
AII					
AIE	150				
	230	230/100	2.3	<50%	
	180				
AFC	100				
AFP	80				
AFS	100 (ong a 35cm)				
CM					
CM					
20 CM	110				
CM					
CM					
10 CM	240	240/100	2.4	±50%	
25 CM	90				
CM					
APOPL					
	90				
	110				
ATA	120				
TRTP	60				
ATP					
AP					

ø 2.1 cm.

L		PSV	PSV^R	RATIO	% STENOSE
AIC		70			
		80			
AII					
AIE		220			
		130			
AFC		160			
AFP		210			
AFS		~			
CM		~			
CM		~			
CM		~			
CM		~			
CM		~			
CM		~			
CM		~			
CM		~			
APOPL		15			
ATA		9.			
TRTP		20			
ATP					
AP					

40
35
30
25
20
15
10
5
0

Opmerkingen en conclusie: - losse mandmogelnytjheden
Re: AIE stene < 50%, AFS stene ± 50%.
Li: AFS geoccludeerd.

9417053 (7-95)

Duplex parameters have gained their meaning by submission to angiography. The calibration of duplex as representational device depended on taking angiography as the gold standard. Only after such a calibration has been put into place is it possible to conclude, in any specific patient, that the duplex and the angiography give either the same result—or one that is different.

In the weekly meeting the case of Mrs. Veger is presented by a radiologist. His hand waves vaguely in the direction of a video screen above him, which shows a duplex graph. "On the duplex," he says, "she seemed to have a stenosis." Then he moves his hand from up there to the light box in front of him on which several angiographic images are suspended. "In reality," he then continues, "she appeared to have this vascular system. Instead of a single stenosis, these pictures show a lot of grave irregularities in the width of the vessel lumen."

The duplex and the angiography of a single patient, Mrs. Veger, say different things. It is like the pressure measurement and the complaints we came across earlier. If they differ, the radiologist takes it that they cannot both be right. He makes the angiography win. There's no argument about which technique might be right and why. There is no explanation that explains away the results of one technique. In this case, the hierarchy is blunter. It is a matter of representational power. The duplex makes things "seem." The angiographic pictures show "reality."

Hierarchies between representational devices may shift in character over time. The radiologist just quoted wasn't very familiar with the new technique. But since duplex is used more and more, it gradually becomes harder to wave a hand at it and say it makes things "seem." On the very day I noted the ex-

and informs the other while they are also different identities. They are not different places the person walks between or can take refuge in. Neither are they alternating facades or two sides engaged in a dialogue. Not two different persons or one person divided into two. But they are partially connected, more than one, and less than many (Strathern 1991, 35). *More than one and less than many.* There it is: in the literature. It is there already! The very image that here, in this book, I try to sketch (give flesh to, develop, color) when talking about the reality of atherosclerosis.

The Organism
Relating to the literature helps to give words backgrounds. A history. Points of contrast. If you have read this subtext so far this may help you to situate the double move made above: to study the multiplication of a single disease *and* the coordination of this multitude into singularity. It should also help you to appreciate why I do not talk of *systems, discourses, paradigms,* or *cultures* when talking about medicine. These terms, however different, all somehow resonate the image of the organism as a model for what it is to hang together.

ample just quoted I witnessed another scene. It shows that even then angiography wasn't always on the top of the hierarchy.

The same meeting, ten minutes later. The case of Mrs. Takens who had an operation six months earlier. Since then her bypass has clogged up again. But how far exactly? It might be occluded, for the angiographic picture shows no dye beyond a critical point: the white stops abruptly. The duplex, however, still shows a peaking graph below this point. Flow. One of the radiology residents asks: "In a case like this, when the angio says 'closed' and the duplex says 'open,' what should one believe?" Two surgeons, speaking with a single voice, say: "Duplex." And then one of them tells how he once studied seventeen cases like this: patients whose angiography showed an occlusion while their duplex showed flow. In all seventeen cases duplex proved to be in line with the findings on operation. "It was only seventeen cases, so I couldn't publish it. But there were no exceptions."

The two surgeons who speak up for duplex here have done a lot of research on the technique. So much research that they are able to make it win sometimes — like in those cases where angiography shows an occlusion and duplex doesn't. An arbiter is cited that makes the duplex win: it is a surgical reality par excellence. It is the reality of the arteries that become visible once a patient's body is anesthetized and opened up with knives in an operation. The blood vessels that the surgeon can see from the inside with his naked eyes — so long as there is no blood flowing through them.

Coordination

If we no longer presume "disease" to be a universal object hidden under *the* body's skin, but make the praxiographic shift to studying bodies and diseases while they are being enacted in daily hospital practices, multiplication follows. In practice a disease, atherosclerosis, is no longer *one*. Followed while being enacted atherosclerosis multiplies — for practices are many. But the ontology

What is it to hang together? In more recent literatures (but how to name all relevant titles, they are so many?) there are other images around. Of clashes that bind. Of coming to celebrate in the same ancestral house or of writing in the same journals. Of engaging in practices that *make* connectedness. Of making translations that draw together *and* establish difference at the same time. There are images around of the patchwork, the fractal, the land-scape, the mixture. And there are blanks: what it is to hang together is turned into an open question. The question of how objects, subjects, situations, and events are differentiated into separate elements and how they are coordinated together is opened up for study.

Is thus the image of the organism left behind? Maybe something else is happening. Maybe this image, too, is altering. How does the organism hang

that comes with equating what *is* with what *is done* is not of a pluralist kind. The manyfoldedness of objects enacted does not imply their fragmentation. Although atherosclerosis in the hospital comes in different versions, these somehow hang together. A single patient tends to be supplied, if not with a single disease, then at least with a single treatment decision. Clinical findings, pressure measurement, social inquiries, duplex outcomes, and angiographic images are all brought together in the patient's file. Together they support the conclusion to treat invasively — or not to do so. This, then, is what I would like the term *multiple* to convey: that there is manyfoldedness, but not pluralism. In the hospital *the body* (singular) is *multiple* (many). The drawing together of a diversity of objects that go by a single name involves various modes of coordination. In this chapter, a few of these modes of coordination were presented. To summarize.

The first form of coordination on which coherence-in-tension depends is to *add up* test outcomes. It comes in two varieties. One of the forms of addition projects a common object behind the various test outcomes: "the disease." If the projections do not overlap, one of them is made to win. A hierarchy is established and the discrepancy between the tests is explained away. The second form of addition comes with no worries about discrepancies. It does not suggest that tests have a common object. Instead, it takes tests as suggestions for action: one bad test outcome may be a reason to treat; two or three bad test outcomes give more reason to treat.

A second form of coordination is that of the *calibration* of test outcomes. If test outcomes were listened to as if they were each speaking for themselves alone, they might get confined within different paradigms. The question whether different tests say the same thing or rather something different would not be answerable — indeed it could hardly be asked. The possibility to negoti-

together? Physiology still has answers to this question — and is investing into improving them. And so do anatomy, genetics, clinical epidemiology, and all other branches of biomedicine. But it has also become possible to give a new *kind* of answer to this same old question. An anthropological answer. It tells that in the hospital the organism hangs together thanks to the paperwork that travels from one department to the other; the correlation studies that correlate the outcomes of different diagnostic tests; the formulae

and pictures that translate numbers and other data back and forth; the meetings where different specialisms come to agree on the diagnosis and treatment of individual patients. The organism in hospital Z (and other places like it) has gaps and tensions inside it. It hangs together, but not quite as a *whole*. It is more than one and less than many. So where we started out with a society that mimics the organism, what we end up with is an organism that clashes and coheres — just like society.

ate between clinical notes, pressure measurement numbers, duplex graphs, and angiographic images only arises thanks to the correlation studies that actively make them comparable with one another. The threat of incommensurability is countered in practice by establishing common measures. Correlation studies allow for the possibility (never friction free) of *translations*.

Separate Localities

The hospital isn't divided into wings with doors that never open. But it *is* divided into wings. The various atheroscleroses enacted in hospital Z are sometimes coordinated and jointly form a single disease that somehow hangs together. But not always. Sometimes the incoherences between different ways of enacting atherosclerosis aren't smoothed away. They are lived with. But how? That is the topic I will address in this chapter. A chapter about moments when, places where, and more specifically ways in which "the object" atherosclerosis is not coordinated into one, but left incoherent. In practice, sometimes, there are gaps — gaps that may come with clashes, but that do not necessarily do so. And if they do, if the incoherences presented here do give rise to controversies, these tend to be local. They rarely spread — though neither do they go away. In hospital Z, divergence does not necessarily imply either conflict or consensus, for the simple reason that there is not always a necessity to search for common ground. Tensions may also persist in a pacified form.

For a long time, much effort in science studies was devoted to the question of *how* controversies close and *what* makes them do so. Is closure a consequence of finding new facts or are all facts open to endless negotiation? Is closure a matter of solving the logical contradictions between theories or of solving the social conflicts between groups promoting theories? The horizon of the discussion was the presupposition that controversies indeed come to rest at an end point in which differences are settled. This mimicked the closure rhetoric of research publications: these are written as if there were a single reality all should be able to agree on, in the end. However, in the hospital it is easier to trace overt, ongoing incompatibilities. There, the technicity of intervening is more important than the consistency of facts. Incompatibilities don't stop patients getting diagnosed and treated. Work may go on so long as the different parties do not seek to occupy the same spot. So long as they are separated between sites in some sort of *distribution*.

This chapter is about distribution. The metaphor is a spatial one: it allows me to tell how difference isn't necessarily reduced to singularity if different "sites" are kept apart. Here and there. Atherosclerosis in hospital Z is not always smoothly drawn together into a single object: *here* another version of the disease may be enacted than *there*, a little further along. I will follow several such versions, with incoherences between them, to which my informants pointed: these are the more overt ones. I'll also talk about tensions that occasionally give rise to clashes even though they are pacified (and often hardly visible) the rest of the time. And I will point to the possibility of changes in the future landscape of the diagnosis and treatment of atherosclerosis that, if they come about, are likely to do so without quarrels being involved. The *localities* over which the reality of atherosclerosis is distributed may be wings in the hospital building but also boxes in schematic drawings of the disease. And as you get to know some of the intricacies of distribution, you'll also encounter more styles of enacting reality and more variants of atherosclerosis.

Diagnosis and Treatment

Enacting atherosclerosis in the department of pathology is not only a matter of directing the medical gaze at bodily tissues. It is also a matter of touch. The pa-

---

Controversies

Difference may come with attempts to co-ordinate. But it may also take the shape of controversy and conflict. In order to get a better insight into the ways we have come to think about these, it helps, again, to explore the literature. And to go back in time. If you want to join me then we will find that for a long time controversy and conflict were central to many theoretical discussions about *science*. In the subtext to this chapter I will try to present you with the outlines of those discussions. The point is not to cover all their highlights, nor to sum up most of what has happened or been written. The point is again and simply to make some landmarks that help to situate what is done in the main text of this book, and especially the way it deals with scientific conflicts, or better: with their relative rarity in hospital practice.

Controversy was pivotal to the *philosophy of science* in the 1960s and 1970s (for an assemblage of concise and well-argued papers, see Lakatos and Musgrave 1970). Many other questions were asked, but they were all somehow related to this first one: what happens if two scientific theories

thologist takes an amputated leg out of a plastic bag. He searches for the appropriate knife and cuts out pieces of vessel. If it weren't for the gloves he wears, his hands would get dirty. And he's not the only one interacting with physical matter. A technician stains the specimen with fluids. Light passes through lenses and slides before reaching his eyes. Enacting reality involves *manipulations*. And yet in the department of pathology enacting atherosclerosis reaches its apogee when the doctor's eyes see an enlarged vessel wall. When the disease is unveiled and *knowledge* is established.

In the consulting room, enacting atherosclerosis is also to do with knowledge. When vascular surgeons hear patients describe the pain in their legs, or when they feel cold feet or weak pulsations, they *know* atherosclerosis. They write down "intermittent claudication" and the findings of their physical examination in the patient's file. What such knowledge needs to do varies from one place to the other. Pathologists know atherosclerosis in order to *judge* the actions of other doctors. Vascular surgeons know atherosclerosis in order to *plan* their next action. So there are various ways of knowing embedded in various activities. And yet they all have at their center a representation of the object, a diagnosis, a fact that can be written on a form, in a file, in an article. *Data* that may be added up, translated into other data, and, if necessary, travel.

With *interventions* this is different. Therapeutic interventions do not primarily yield facts; they are supposed to *change* the object with which they interact. They must *improve* the patient's condition. They enact an object by altering it. Writing things down in a report afterward is a side issue. Furthermore, intervention reports do not center around a representation of the object intervened in but around the intervention. During diagnosis, talking, touching, cutting, and coloring are the inevitable means necessary in order to gather knowledge; during therapy their material effects are actively sought. The material effects that are

contradict each other? Will *the data* decide which is the better, or is that impossible, since each theory generates its own data and excludes those that fit the other? Another question was whether scientists are faced with conflicting theories as a part of their normal, everyday practice or only in rare revolutionary situations. And maybe, some suggested, *theories* are not the entities weighed against each other as science progresses, maybe choices tend to be made at a higher level, between *research* *programs* of which individual theories form only a part.

In the same period, various *sociologists of science* were no longer satisfied with the study of the social preconditions for and the social effects of scientific knowledge; they began to ask questions about knowledge itself. In doing so they turned *against* philosophy, claiming to be in a better position to understand the sciences. They turned against philosophy, but also depended on it, for they imported the same

minimized in diagnosis form the very aim of therapy. Thus, though after a diagnostic test it is possible to bracket practicalities and concentrate on the object, this doesn't happen with the practicalities of therapy. They *are* the therapy.

Imagine a patient who has an atherosclerosis "that requires invasive treatment." The vascular surgeons have decided to operate. The patient agrees, is admitted to the hospital, forbidden to eat, given sedative medication, rolled to the operation theater, anesthetized, opened up. At that point the surgeons get to see atherosclerotic arteries with their bare eyes. Blood doesn't flow through a leg when its arteries are opened up: the supplying arteries are closed off with a clamp. During an operation the patient feels no pain. So the atherosclerosis in the operating theater doesn't show pressure drop or clinical symptoms. It is, instead, a fleshy affair: an artery with atheromatous plaque inside its lumen that shouldn't be there. This fact isn't revealed in order to represent it. It is revealed so that it can be stripped away.

It is a fat leg. Nurses have colored the inside of the thigh yellow with iodine. The surgeon makes a sharp straight cut that opens up the skin. The fat underneath it is carefully separated by a resident. Blood repeatedly obscures the view. Tissues are used to absorb it. Small vessels are closed off with a hot pin. Larger ones tied off with blue threads. Heparin is added to prevent the blood from clotting. The nurses hand whatever is needed to the surgeons. When things get difficult, the instruments shift almost imperceptibly from the hands of the resident into those of the surgeon. Is this the nerve? Yes, and it is held to one side with a clamp. The entire cut is then widened with a far larger clamp, double, like a pair of scissors. Ah, finally, there is the artery. An orange plastic thread is put round it to mark it. Then a similar search for the artery is repeated just above the knee. Once the artery is made accessible at both places, it is closed off in the groin with another clamp. "There's no more leg," the surgeon warns the anesthesiologist, who keeps a constant eye on the patient's blood pressure.

primordial scene studied by the contemporary philosophers of science: the controversy. The scene in which two theories contradict each other. This move may have been facilitated by the philosopher's habit of casting arguments about logical reasoning in terms of war metaphors (as is so aptly demonstrated in the classical study of Lakoff and Johnson 1979). Sociologists took this metaphor seriously when they interpreted scientific scenes in a novel way. They claimed that controversies were *conflicts*, social conflicts. The question about which of the theories would win was one that was social.

Thus, controversies were no longer moments when *theories* contradict each other. The sociologists took it that only *people* have such capabilities. In the concluding remarks of a case study of a controversy in biology, MacKenzie and Barnes put it like this: "We might say that just as the

I stand on a small stool, right behind the surgeons' backs, and look over their shoulders into the operation area. Like everybody else I wear green clothes and breath through a mask. I have washed my hands endlessly but am forbidden by the nurses to touch anything. It is warm. The surgeon makes two incisions in the vessel wall: one in the thigh and the other further down, above the knee. Then a nurse opens a plastic bag and another one takes a remarkably primitive device from it. It is a ring attached at an angle of 45 degrees to a long, thin, inflexible wire. With a knife the resident loosens the atheromatous plaque from the rest of the arterial wall in the two places where the artery is opened up. He then inserts the ring of the stripper around the plaque. The surgeon feels to check. Right. The stripper is moved upward. Slowly. When it finally arrives in the groin, the entire stretch of atheromatous plaque has been loosened from the rest of the vessel wall. With tweezers the surgeon draws it out. He drops it in a small bin. There goes the thickened intima. With lots of debris attached to it. Its bright white contrasts with the grayish artery that is closed again: patches made out of the walls of one of the patient's veins are inserted in the cut to prevent the artery from forming a stenosis where it is stitched up.

The action undertaken here is called endarterectomy. The disease enacted is an encroachment of the leg arteries. There are knives that cut, clamps that prevent blood from flowing, skillfully moving hands. There are drugs, breathing movements of a lung machine. Together these actors lay the artery bare and make two holes in it: one high up, another further down. And then there's a ring stripper. Enacting atherosclerosis with a ring stripper is not a matter of unveiling it, but of stripping it away.

Multiplicity is complicated. Not only are there different "atheroscleroses" enacted in any single hospital, but there are also different styles of enacting these. There is diagnosis, in which the questions "what is the matter?" and "what to

two communities [in the case studied] collectively 'decided' to remain separate, so they collectively 'decided' to define their theories as incompatible. They were not forced to do this by any 'inner logic' " (MacKenzie and Barnes 1979). What looks like a *logical contradiction* to those who follow what MacKenzie and Barnes call "formal modes of thought" is described in different terms by MacKenzie and Barnes themselves. They tell about "two communities" that are separate from one another. The two communities might "decide" to fuse or otherwise engage into a compromise. But they don't. They present their ideas as contradictory as a way of engaging in a *conflict.*

Where philosophers of science were concerned with the way logical contradictions were handled in practice, these sociologists said that "logical" contradictions do not exist outside the practice in which they are defined. Social conflicts *generate* contradictions. The term *controversy* allowed a subtle sliding movement from logic to sociology (and, if need be,

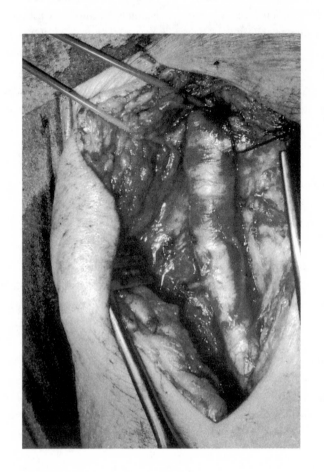

do?" alternate and intertwine. And there is treatment. In treatment, doing is a matter of undoing. *Enacting* disease takes the form of *counteracting* it. But however much these styles of engaging with reality differ, the object, the "atherosclerosis" that is treated, *may* be similar to the "atherosclerosis" that was diagnosed earlier on.

On one of the walls of the operating theater a light box hangs. The surgeon who is about to operate looks from the leg on the table to the X-ray pictures hanging on the light box. Yellow colored skin here, black-and-white shadows there. "Look," he says to me, pointing with his gloved finger to one of the images, "that's the stenosis. Do you see it? It's easy to see, here, look." Then he moves the same gloved hand to the leg below. "Now we'll make the high cut here. And then the low one there. And in between those places we'll strip the lumen clear." His hand points at a "here" and a "there" twice. On the light box, in the leg. As if he were indicating places that correspond. Places on the black-and-white image and in the fleshy vessel that are the same.

Angiograms are two-dimensional plastic pictures smelling of fixer, and the atheromatous plaque stripped out of arteries in the operating theater is a twisting, bloody structure. So they are not the same. But they enact atherosclerosis similarly as an obstruction of the vessel lumen. The distance between them is small enough to be bridged by a moving finger.

The vascular surgeons of hospital Z use angiographic pictures as maps during operations. The images help them to decide where and how to cut. Thus, angiograms are part of the diagnosis: they help to decide how and where to operate. But they do not play a crucial role in the decision about whether or not to treat. This decision has already been made earlier on. The surgeons of hospital Z only ask the radiologist to make angiograms once they have decided that an operation will be done. In hospital Z, bad pictures are not used as a reason to treat. But bad complaints, bad ankle pressure, and bad troubles are.

back again). In the late seventies and throughout the eighties a lot of *controversy studies* were done by sociologists studying science. They depicted the closure of controversies as a social phenomenon. Something that depends on power, force, numbers. Whatever. Reason is never decisive, the reasonable is an outcome. The closure of a controversy means (or so the sociologists concerned took a lot of effort to show) that one truth wins and the arguments in its favor retrospectively become those that are reasonable.

But not all scholars who engaged in the study of controversies in science submitted "formal thinking" to social relations. Many were able to keep logic and the conflicting interests of social groups apart. An example. In their introduction to the edited volume *Scientific Controversies*, Tristam Engelhardt and Arthur Caplan make a fascinating *list* of possible causes of clo-

Vascular surgeon: "In some hospitals they decide whether or not to operate using X-rays.
They go by the percentage of lumen left. We don't. For us an angiogram is more like a road
map. A photo makes it easier to approach the vessel. But we treat patients, not pictures.
Sometimes we get X-rays from a peripheral hospital. And they say, look, isn't that bad.
Something must be done, it looks difficult, please, could you do it for us. But of course we
always talk with the patient first. And we only operate if that patient has severe complaints.
And they must be pretty motivated too, patients, before we operate."

In hospital Z the decision whether or not to operate doesn't depend on angiography. Sure, if no stenosis were visible on the angiogram, the surgeons wouldn't know where to intervene. But before an angiogram is made in the first place, a patient must have bad complaints, be seriously hampered by them, and be motivated enough to incur the risks of an operation. So the disease diagnosed *may* be the same as the disease treated, but this *doesn't need* to be the case. In hospital Z it isn't. The atherosclerosis enacted in the process of deciding that an operative intervention will be done differs from the atherosclerosis enacted during the operation. "Pain when walking" is the *reason to* intervene, whereas "a plaque that encroaches the vessel lumen" is the *target of* an operative intervention.

This is an incompatibility. The disease diagnosed and the disease treated are different objects that, in ways we have come across earlier, may coincide but do not always do so. If such an incompatibility would be considered disturbing, it would not be necessary to still live with it. There are possibilities for aligning the objects of diagnosis and intervention. A first one is that of making "vessel obstruction" dominant. In that case a stenosis on an X-ray image would become the reason for treatment, the obstructed vessel lumen found in the operating room the treatment's target. The second possibility is to make "pain on walking" dominant. In that case, operative treatment would be banned. Instead "pain on walking" would become the target of treatment as well as the reason to treat. A

sure: (1) closure through loss of interest; (2) closure through force; (3) closure through consensus; (4) closure through sound argument; and (5) closure through negotiation (Engelhardt and Caplan 1987, 14–15). A list. As if these five possibilities exist side by side, next to one another. A theoretical dispute is avoided. Instead, there seem to be various empirical possibilities. To name just the extremes: some-

times controversies close because there are sound arguments in favor of one of the theories at stake—and sometimes they close because one of the social groups involved proves to be the strongest.

Simply listing these closure mechanisms as ever so many possibilities in a calm and neutral way, however, misses out on everything that was at stake in the theoretical dispute. To begin with, it

treatment practice directly intervening in "pain on walking" exists: it is called walking therapy.

Angiologist: "Now that you mention it, yes. Yes, I know about walking therapy. Of course. It's in the literature. The literature tends to be favorable. Impressive clinical trials. Yes." Interviewer: "Then why isn't it done here?" Angiologist: "Well. You mustn't forget that before we see them, general practitioners have often encouraged these patients to walk. With no particular result." Interviewer: "But that's not the same, is it?" Angiologist: "Well, maybe not. Yes. No, it isn't. It's likely that walking therapy only works if there's some structure to it, good supervision. Okay. We haven't organized that here. Why not? I think because it would be something for the physical therapists. And they've had budget cuts. Whatever you ask them these days, they just say sorry. They're not allowed to do more, they must reduce their activities. So they won't jump into new adventures."

Walking therapy is a treatment strategy that requires patients to walk for long stretches regularly. There is a variant that says that patients must keep on walking even when their legs hurt. Another allows patients to rest for a while just before pain is to be expected. There is a version that uses an indoor treadmill. Another encourages people to go outdoors. But whatever the precise shape it takes, all walking therapy fights an atherosclerosis that hurts when walking. *By walking.*

Coherence could thus be strived after. Try to establish the nature of atherosclerosis, the principles of the disease. And organize practice accordingly. Say, for instance, that atherosclerosis *is* legs that hurt on walking. And base your practice on that: diagnose this disease in the consulting room by talking and treat it by means of long outdoor walks. Or establish other principles and say, instead, that atherosclerosis *is* an obstruction of the vessel lumen. Build a prac-

misses out on the political message of the sociologists. When sociologists of science claimed that social conflicts precede and generate logical arguments, it was important to them to stress that this is not just one option from a list of five. Only if the closure of a scientific controversy *always* depends on social factors may the "scientific controversy" come to be recognized as an intrinsically social event. And only then does sociology have a novel message. This message: current scientific theories do not depend on capturing na-

ture by reason because they are not the only "reasonable" theories possible. Alternatives existed until very recently. They were subscribed to by respectable groups of people. So the problem with these alternatives is not that they were (or that they are) unreasonable, but that they have lost.

The social message implied was that present-day experts do not represent "reason," but happen to inherit the ideas of those earlier experts who have managed to dominate their rivals. This means that societies who hand too much power to their ex-

tice on that: diagnose the disease through images and treat it by surgery. But in the hospital, in hospital Z, it doesn't work like that. Such coherence isn't achieved—or even sought. Practice isn't preceded by a principled discussion. There are, instead, different practices that each contain principles of their own. Apparently what seem to be logical incompatibilities are not disturbing. They don't make life more difficult: they make it easier. Atherosclerosis in hospital Z *is* one thing here and it *is* something different a little further along. It is *both* pain *and* a clogged up artery but not both in the same site. It is pain in diagnosis and a clogged-up artery in treatment. Reality is *distributed*.

Indication Criteria

Treating atherosclerosis by stripping it away in the operation theater is not the only invasive treatment available to the vascular surgeons in hospital Z. There are others. There is, for instance, percutaneous transluminal angioplasty, PTA. Although the surgeons are responsible for all treatment decisions, this treatment is not one they do themselves. It is done by radiologists in the invasive room of the department of radiology. The room where angiograms are also made.

A patient wearing only an orange shirt lies on a high table in the radiology department. A large X-ray apparatus hangs over the sedated body. Around the table, above our heads, are monitors. They show vague X-ray shadows of the patient's arteries, which contain a low dose of dye. The catheter, entirely opaque, is visible, too. With his eyes on the monitor the radiology resident slowly pushes the catheter from a hole inside the patient's groin down into her femoral artery. Somewhere along the catheter, somewhat below its tip, two dots are

perts have a problem. This is caused by the fact that experts are not necessarily neutral in relation to the conflicts that they are asked to solve: they are more likely to be a part of them. (For a change, I will relate here to *a journal* instead of a book. *Social Studies of Science* has a long history of publishing studies that analyze controversies in a neutral tone, equally suspicious of all experts.)

There was also a more radical message drawn from the partiality of expertise. It was held by those who hoped for a completely different, better, alternative society.

These were not neutral in the way they talked about different experts (so as to avoid siding with the winners) but sought out those forms of expertise that linked up best with their aspirations for a just society. They were trying to find, invent, invest in, an alternative science. Their argument was that society can only be changed if the science we use to build it is changed as well. And, vice versa, an alternative science will only emerge as a part of different social relations. Since a society and its scientific products are intertwined, changing one will be a process that also involves

visible. At a certain moment one of them is at each end of the stenosis. "Look, there's our balloon, neatly where we want it," the radiologist says.

A second resident attaches a small pump to the line. The way he pumps reminds me of inflating bicycle tires with air. Between the two dots a balloon expands. It stretches the vessel walls. The balloon is kept inflated for a while. "Does it hurt, Mrs. Zenga?" one of the residents asks, in a voice much louder than he had used a little earlier while talking to his colleagues. "You have to warn us if it hurts. Not if it just feels itchy. But if it hurts." The sedated patient looks invaded. Absent. Old. "I think that's about enough, how long have we been holding it now?" the radiologist asks his residents. The air is allowed to escape. And someone says: "I guess this will do. Let's inject some contrast. Let's see how large this lumen is."

Like endarterectomy, this intervention, PTA, enacts atherosclerosis as a plaque encroaching a blood vessel. But when it is stripped away, atherosclerosis is *taken out* of the body. It ends up in a container. Whereas a plaque that is thrust sideward with a balloon stays inside the body. It is *pushed aside.*

There's yet a third important invasive technique available in hospital Z. It again depends on opening up a leg with surgical knives in the operating theater. But, unlike both endarterectomy and angioplasty, it doesn't "unplug" the vessel lumen at the site of the encroachment. Instead it offers the blood a way of bypassing the stenosis, of flowing around it. Upstream from the place where the lumen is encroached a bypass is attached. Downstream another attachment is made. Thus, a bypass enacts atherosclerosis as something that may be *circumvented.*

The upper right leg has the stenosis. The lower left leg, however, is opened up first. A vein is taken out. Its valves are carefully destroyed. All of them, for if they were left they would impede blood flow and be a good place for new plaque to attach itself. "A lot of work," the

changing the other. (See for this the various volumes of the late *Radical Science Journal*.)

Both these messages, the warning against expertise and the plea for an alternative science, get lost when "reason" and "force" (or other social mechanisms such as consensus and negotiation) are listed alongside each other as equally plausible mechanisms for the closure of controversies. Such a listing suggests that there is a phenomenon called "reason" that is an independent asset, a nonsocial phenomenon, and that this "reason" is *a way out* of the social. A good clear way out of the mess of the social—which is how philosophers of that time depicted reason. As defenders of "reason" they never said that closure through force is not an empirical possibility. It may occur. But it is not good. The knowledge that results from forcefully closed controversies just does not deserve to be called "scientific." This points to what was at stake for the philosophers, some-

surgeon tells me, "and it's quite something to cut into two legs. For us as well as for the patient. But the results are much better than with artificial bypasses. We use those, too. Sometimes you have no choice. But they clog up earlier. Veins are better. They're not a foreign body." The upper right leg is opened, the artery laid bare. Not at two spots now, but over the entire stretch that the bypass will circumvent. Upstream from the stenosis a small hole is cut in the artery and the bypass is sewed onto it with small stitches. And this is also done downstream. With its extra piece of vessel in it, the leg is then closed again. Thread, needles. Tweezers to pull the needles with. The skin is stiffish.

Three invasive treatments that all enact atherosclerosis as a stenosis in the arteries, but in slightly different ways. Endarterectomy strips the vessel encroachment away, angioplasty pushes it aside, and a bypass circumvents it. A patient doesn't need all three treatments, one piled on top of the other. A single one will do. A single one will have to do. But which one? One might imagine a controversy about this question. A controversy during which the question "which of the three treatment strategies is best" would be solved. After that controversy atherosclerosis would be enacted in one of three ways: it would be stripped away, pushed aside, or circumvented.

During my fieldwork I came across instances of just such a controversy about invasive treatment. The complete eradication of each treatment strategy was predicted by a spokesperson of another.

In a large lecture room of the medical school linked up with hospital Z there is a small international conference about PTA *and other so-called endovascular procedures. The last speaker is a vascular surgeon. He announces that this professional attachment gives him the authority to announce the end of vascular surgery. "It will come about quite soon. Endovascular procedures are winning. There are still a few problems to solve. But since* PTA *does not require a major operation, far fewer patients are likely to die during treatment. It is the*

thing that the Engelhardt-Caplan list again skips over—just as lightly as it skips over the political message of the sociologists. This, then, is message number two that gets lost. When defending "reason," philosophers were not making an empirical claim but were setting a normative standard. They did not say social powers never closed a controversy, but that this was dangerous. Scientific controversies *should be* settled through "sound arguments." If other factors interfered, then the realm of science was being abandoned in favor of that of politics. And it was important to keep the two apart. Only thus could science (and the rights of reason) be protected against the arbitrariness of social powers.

An example of why this was so important was provided by the Lyssenko affair. The Lamarkian biology of Lyssenko had acquired support in the Soviet Union because it fitted nicely with the locally dominant political ideology. It said, to say it

more elegant technique. So whatever problems there are, they will be solved. And vascular surgery will become obsolete. Mark my words."

I marked his words down in my notebook. And this allows me to contrast them here with the quite different words that I noted down a few months earlier. These do not announce the forthcoming victory of endovascular treatment, but of the other protagonist, vascular surgery.

We are on our way to the elevator. It's an ordinary Monday afternoon, just after the decision-making meeting. One of the vascular surgeons talks to the angiologist. "If you ask me, I am starting to have doubts about this whole business of PTA. There are so many problems with it. Arteries closing off again. Side effects like a thrombosis further down. And I've come across several negative reports in the literature recently. Maybe it isn't such a good technique at all. Maybe we should do some bookkeeping. I doubt whether after that we'll still want to go on with it, after some proper bookkeeping."

Whereas one surgeon announces the eradication of all surgical treatment in the near future, the other says that proper bookkeeping will wipe out PTA. The controversy looks as fierce as can be. But while such controversial statements are made at conferences and in informal conversations, the very people who make them keep on prescribing surgery as well as PTA in their daily hospital practice. So if a full-blown controversy would imply that a technique would be likely to indeed disappear, small instances of controversy don't necessarily do. They may also point at a tension that lingers on for ages. All the while the techniques keep on coexisting. They are distributed.

In daily hospital practices a lot of energy is spent on the distribution of treat-

---

quickly, that plants are capable of adapting to their surroundings and of passing the adapted traits on to their offspring. Large-scale agricultural projects were founded on these theories. The "false" Lamarkian "facts" were imposed by force on the biological community and the collective farms of the Soviet Union. But such an imposition of politics on biological reasoning (or so the argument went) cannot but have disastrous effects. And so it did. Lots of people died of starvation when one harvest after another proved a failure. (For a version of this history that tries to save both

science and Marxism by directing all the blame at Stalinism, see Lecourt 1976.)

If Engelhardt and Caplan make a list of five possible ways to close a controversy, without engaging with their relative value or their relations, they *distribute* theoretical options over empirical cases. (This struck me from the first moment I saw it since *distribution* is a rare way of solving, or rather dissolving, theoretical tensions inside the texts of theorists, whereas it is a common and routine way of doing so in more practical settings such as the hospital. This gives you another link between hospital Z

ments over patients. Instead of the question "is PTA good *in general*" the mundane daily question is "is PTA good for *this specific patient*." This is the kind of question that daily negotiations focus on.

A senior surgeon addressing a junior, in front of a light box on which an angiogram is suspended, pointing at it: "What, do you want to propose a PTA in this patient? Are you crazy? Come on, that won't be any good, it's almost occluded, here, this bit. They'll never get a catheter through that."

In opposing a PTA in an artery that he takes to be occluded this senior surgeon is outspoken. Severe. But the quarrel is local. It concerns a single patient. And even if there may be disagreement about single patients, quite often there isn't. Quite often there are no clashing securities, but insecure doctors who face the question "what to do?" They hesitate.

The decision-making meeting. A junior surgeon presents a patient, Mr. Lethaman. He's got serious complaints. His ankle/arm index is 0.6. The duplex shows a quite long stenosis, there it is, it seems to be about ten centimeters long. In the superficial femoral artery left. What might be wise? No, it's not a complete closure, and he thinks it isn't necessary to go for a bypass. But is this lesion a suitable candidate for angioplasty or is it better to put Mr. Lethaman on the list for endarterectomy? Various people come with arguments for one or the other and the treating surgeon nods at all of them. They've all got a point—so how to come to a conclusion?

How to treat Mr. Lethaman? The surgeon who is responsible presents this as a question to his colleagues of the vascular team: surgeons, radiologists, and the angiologist. He doesn't have a standpoint to defend in a controversy, for he isn't quite sure what to do. Which therapy might best be suited for his patient?

and a text out of the literature I relate to here: an *analogy* between their ways of handling difference.) But what is hidden in this pacifying gesture? The question of how to think about and work on the relation between science and politics. Is it wise to try to protect science from politics *or* is it a dangerous illusion to believe that scientific experts are politically neutral? Should the powers of the day be prevented from intruding on science *or* are experts a new and powerful social group who deserve to be put under some form of democratic control? Should the sciences be left to run according to their own self-determined rules, or have the boundaries between the inside and the outside always been leaky, keeping apart some elements and issues, but not all of them.

And, with these, there is the question of what kind of politics to engage in: one of setting standards or another that, convinced of the messiness of the nonconforming world we live in, seeks better ways of handling it.

The question of which treatment might be good for a specific patient hasn't disappeared after a decisive controversy that made one treatment win over the others. But neither is it a completely local question again and again. Instead, the distribution of treatments over patients is facilitated by *indication criteria*. Indication criteria are distributive tools. They link up patient characteristics to one of the available treatment strategies. In hospital Z, indication criteria are part of the scientific output of the vascular team. The literature is digested, rephrased, and supplemented. Here's an example. In "lesions longer than ten centimeters" PTA should not be used.

"Studies by Zeitler et al. and Capek and colleagues have shown that the failure rate of PTA increases dramatically if lesions > 10 cm are treated; PTA should not be used in such cases" (van der Heijden FH, Eikelboom BC, Banga JD, Mali WP: The management of superficial femoral artery occlusive disease. British Journal of Surgery 1993; 80:959–96).

Indication criteria mobilize disease characteristics that are easy to establish. The length of a stenosis, for example. The one quoted here turns ten centimeters into a cutoff point. In lesions longer than ten centimeters PTA fails so often that it isn't worthwhile. But in lesions shorter than ten centimeters PTA also has a higher failure rate than endarterectomy. And yet it is done. Why?

"Although the patency of endarterectomy is considerably better than of PTA, PTA is a non-surgical technique with a relatively low morbidity rate. Furthermore patients need only a short hospital stay" (van der Heijden FH, Eikelboom BC, Banga JD, Mali WP: The management of superficial femoral artery occlusive disease. British Journal of Surgery 1993; 80:959–96).

Arteries are more likely to stay open for a while after having been stripped than after having been inflated. But people are much sicker from operations than from angioplasty. They also need to stay in the hospital far longer. And they run a greater risk of death. So the calculus that supports the distribution of

Nonclosure

In 1984 Frances McCrea and Gerald Markle published an interesting article about estrogen replacement therapy (ERT). Scientists in the United States and in Britain in the early eighties, or so McCrea and Markle tell us, give different advice about this therapy. "Most US researchers do not recommend the therapy, claiming that it increases the risk of cancer; in Great Brit-ain, however, most researchers minimize the cancer link and advocate ERT for conditions ranging from hot flushes to osteoporosis (loss of bone mass)" (McCrea and Markle 1984, 1). So there is a difference. But there is no controversy. Even in the absence of a language barrier researchers on the two sides of the Atlantic managed to not relate much to each others' research publications. They did not fight. Neither

treatments over patients is heterogeneous. It contains the length of lesions as well as that of hospital stays. It contains diagnostic facts as well as practicalities of diagnosis and treatment.

In treatment practices, atherosclerosis is not turned into a singular reality, either something to be circumvented, or something to scrape away or something to push aside. It is all three of these realities. But not all three at once. They are distributed over different groups of patients by means of indication criteria. For every patient group, another treatment, and thus another disease, is indicated. The indication criteria are a place where it becomes visible that "reality" may inform practice while "pragmatics" in their turn shape reality. The two are interdependent. If an operation is preferred *because* it effectively scrapes away or bypasses the disease, then practical arrangements follow the reality of an "encroached vessel." The patient is put under anesthesia, cut in, and operated. But if PTA is chosen *because* it entails less risks and is a smaller burden on the patient, then these pragmatics dominate. Reality follows: atherosclerosis is enacted as something that can be pushed aside by a balloon.

Stages and Layers

Vascular surgeons treat people who are in a bad condition. What is their disease? Their legs hurt. Their skin looks bad. Their ankle pressures are low. Their blood velocity is locally increased. Their angiograms show a stenosis. Their arteries contain plaque. So far I've stressed that there are differences between the atheroscleroses enacted in each of these cases. But they also have something in common. In vascular surgery atherosclerosis is enacted as a *condition*. It is a problem here and now that plagues a patient and may be treated invasively or not. But there are other wings in hospital Z. In the department of internal medicine atherosclerosis is a *process*.

did practicing physicians—who also disagreed. Practitioners in the United States prescribed lots of estrogen and those in Britain didn't. The authors present a further difference: feminists and health activists in the United States resisted the prescription of estrogens for menopause by their physicians, whereas those in Britain demanded it from doctors who were unwilling to comply.

McCrea and Markle tried to explain the various positions in these noncontro- versies by linking them to the situations the various groups find themselves in in the two countries. Most American researchers in this field are publicly funded, and most British researchers are funded by the pharmaceutical industry. American physicians earn money from their estrogen prescriptions, but for British physicians they simply imply extra work. The American feminists have taken "biology as destiny" as their long-standing enemy, whereas the British feminists attribute the

The senior internist has a research program going that investigates the relation between lipoproteins in the blood and the development of atherosclerosis. He's critical of vascular surgery. He thinks that what vascular surgeons do, even if it is done well, isn't always a good thing to do. "They prefer to neglect risk factors. They pretend these do not matter. Pipes open again? Fine, then the patient is saved. And they portray themselves as the heroes who've managed to do the saving. I'll never forget this picture in the newspaper. It's old now, but it's still significant. The picture was of a man who had had one of the first heart transplants. There he sits in his hospital bed, three days after the operation. He still must have felt awful, but he smiled because of the photographer. And then there's the caption that says he's well again, for, look, he has his usual breakfast: bacon and eggs."

Bacon and eggs are bad for the fat balances in the blood. They lead to the development of atherosclerosis. They clog up the pipes. So it may seem heroic to transplant hearts, and those who are able to do this may pose as lifesavers. They have opened up the pipes. But as long as they allow their patients to eat bacon and eggs for breakfast, the pipes keep clogging up. And the surgeons involved deserve to be unmasked as killers.

"Killer" is a strong word. None of my informants ever got near to using it. But the interview just quoted contains all the intellectual ingredients necessary to back up such an accusation. In theory, the clash is full blown. In practice, however, it is, once again, more complicated.

Someone has told me there's a young internist who's undertaken a study that might be of interest to me, for it concerns atherosclerosis. So when I meet him in the corridor I ask him whether he's willing to give me some of his time. "Of course you can come and talk with me. Be my guest. But there's very little I can tell you about my research. I mean: I'm asked to

disadvantaged position of women to structural causes and see "neglect" as part of this. Thus *social explanations* for all positions are provided. "The positions of various interests, and their systematic opposition to one another, are seen as outcomes of political, ideological and economic relations" (18). But that it contains social explanations is not the reason I relate to this article here. In the early eighties many social studies of science publications referred to "political, ideological, and economic relations" while talking about science and medicine.

The surprising thing is that McCrea and Markle do not talk about parties being engaged in controversy until it comes to a closure. Instead, the image is that of a series of juxtapositions, with different frictions that do not move towards a resolution. Maintaining distance, or so McCrea and Markle argue, is even instrumental for some of the parties concerned. For instance, for the feminists, whose strategy is one of political opposition. "Thus women in both countries tried to neutralize the stigma of menopause: in the US, feminists tried to neutralize the stigma by claiming

unravel links between a fairly rare lipoprotein disorder and atherosclerosis. But so far I have too few patients to be able to say anything sensible about it. I don't get patients. The point is: they only have very few symptoms from their lipoprotein disorder. Their main symptom is the early onset and fast development of atherosclerosis. Which means that they do come to the hospital, but they don't come to see me. They see a vascular surgeon, or a neurologist, or a cardiologist. And I don't even know whether their lipoprotein levels get measured. If they are, the patients are maybe sent to a dietitian or prescribed a drug. What do I know? All I know is that they don't send them on to me. So I'm a bit stuck, really."

It is difficult to raise a controversy about the treatment of patients with lipoprotein disorders if one doesn't get to see the patients who are suffering from it. It is difficult to raise a controversy with vascular surgeons if they don't refer the patients concerned to the department of internal medicine. So however worrying it is for internists that surgeons intervene in the condition *encroached vessels* while they neglect the process of *vessel encroachment,* they are not in the position to raise a controversy.

Here, atherosclerosis is enacted as a present condition, there, as a process that has a history. Tensions between these ways to enact the reality of the disease are articulated. But it doesn't come to a full-blown fight. Instead, the differences between the condition atherosclerosis and the atherosclerotic process are distributed. I might describe this distribution as one over specialisms: in vascular surgery atherosclerosis is enacted as an encroachment of the arteries and in internal medicine as a process of encroachment. But it is more complicated. These two specialties do not just distribute each others' reality away. They do not just push it elsewhere, outside, but they also create a place for it. They create a place for the other reality inside their own—on their own terms. Thus,

that menopause is normal and not a disease; in Britain, women claim that menopausal problems are "real" and not just in their heads" (16). Differences aren't necessarily bridged: they may be kept open—with suitable hard work. They need not be overcome, be it by agreement or force, they may just keep going. Here, then, we have tensions that do not come to a conclusion. In the United States as well as in Britain feminists are angry at physicians and cite researchers. There is no single place, however, where everything comes

together. No single "theory" wins—and neither are the opposing ones fused into a compromise.

This wasn't stressed by the authors, and it became a topic *in the literature* only a lot later. But in retrospect the article just quoted makes an intriguing move. It suggests that the primordial scene studied by philosophers and sociologists alike, that of a controversy that comes to a closure (whether by reason or by force), may well be a rare event, not because it is only in "revolutionary" situations that things do

the condition "vessel encroachment" becomes a late *stage* in the process of encroachment, whereas the "process of encroachment" becomes an underlying *layer* in a patient's poor condition.

What does such a complex distribution look like in practice? Let's move to the outpatient clinic of internal medicine. In their outpatient clinic, internists don't fight surgeons, they fight atherosclerosis. They try to reduce the progression of the atherosclerotic process in patients at risk. In as far as their preventive measures fail, the atherosclerotic process develops, which it may do until the patient has claudication complaints and vessel encroachment. So in the internal medicine outpatient clinic, a poor vascular condition is enacted as the result of a failure to intervene earlier on. If, but only if, it happens to come about, vascular surgery is called for.

The internal medicine outpatient clinic. I sit in with an internist who is seeing patients with diabetes. The consulting room looks like the other one, a few corridors away, where the surgeons work. Small rooms with no windows. The waiting rooms have windows. (The architects made a lot of noise about this: patients deserve windows.) A woman in her early thirties comes in. "Ah, there you are Mrs. Dam. There you are, welcome. We have a guest today, she investigates doctors. She looks at all I'm doing. Well, how are you doing?" Mrs. Dam is doing fine, she says. But she's curious. She felt fine last time as well. But her lab results were bad then. What about them now? The internist nods. "I've got them here, look. They're a lot better. Better than ever, I believe. How did you do this?" They jointly look down at a list of numbers produced in the hematological routine lab. Sugar level. Total cholesterol. Proportions of high-density and low-density lipoproteins. And they discuss Mrs. Dam's attempts to keep her blood levels under control. "It may be that this time I lived as regular a life as I could. I did. No more night life. All meals at the same time every day. Proper meals. And no shifts in the hour of my insulin injection. I went mad. But I thought: I have to be able to do it."

not fit, but rather because they hardly ever do. If one shifts one's scope a bit, from an apparently smooth outcome to a disturbing detail, or from a small site to one that is larger, things tend to get more complicated. If one doesn't simply focus on research in the United States but includes British research as well, divergence becomes visible. Or if it isn't only researchers that are taken into account, but practitioners and then activists as well, the picture

keeps on shifting. There may be differences without conflict. And there may be conflicts that never come to a conclusion.

In political theory it is an old trope: unresolved conflicts come as no surprise. There are shelves filled with volumes on topics such as how wars are fought and how they are avoided; how democracy came into being in some parts of the world as a way of handling difference; how difference was handled differently in other places; how

In the department of internal medicine atherosclerosis doesn't have to do with complaints, but with the future. Current blood levels are measured. Patients with diabetes are prone to get atherosclerosis. But the better their sugar and lipoproteins levels behave, the lower the chances of atherosclerosis progressing (or so the most up-to-date clinical trials suggest). Thus, effective control of the blood now may avoid the need for operations in the future. If the process is monitored carefully from the very beginning, atherosclerosis may never reach the point where it causes trouble. But this isn't certain. Things aren't always as they should be. Some bodies do strange things. Some patients do strange things. And internists sometimes fail as well. Then surgery is called for. If atherosclerosis is enacted as a process, then it is located in time. And somewhere in the future, further on along the time axis, if things don't work out well, a poor condition may develop. One that requires surgical treatment.

In the department of vascular surgery, enacting atherosclerosis as a poor condition leaves room for its process of development as well. Vascular surgeons don't fight the measurement of blood levels of lipoproteins and sugar. They even practice it. In young patients, for instance. If young patients have developed atherosclerosis, the blood levels of their lipoprotein may be deviant. If so, they should not only be treated surgically, but also be referred to the department of internal medicine. Without a simultaneous intervention in the underlying process, the patient is most likely to come back again in a few months, with new complaints.

A vascular surgeon: "What did you say? Mr. Jenner? But in a patient like that we've done blood measurements. I'm sure we have. We always do blood measurements if people are young. Now what young means, it may depend. But this man, what's his age, forty-five or something. And you can't find anything in his file? Very strange. It must have been done. I'll look into it."

various sorts of oppressions are lived with in various ways (from long before Moore 1966 to after Benhabib 1996). I cannot hope to make a serious link to the literature of that tradition here. But just to illustrate that it might be made, here's a small, esoteric example. In 1968 political theorist Lijphart published a book about the way difference was handled in Dutch political culture from 1917 up to 1967. Differences were pacified. Accommodated. Dutch social life (from political participation to playing sports) was organized in several coexisting, nonoverlapping communities (Protestants [of different denominations], Roman Catholics, liberals, and socialists). Dutch pluralism took the shape of a division of the population into *pillars*. Those on top of the pillars, the elites, met each other in parliament and several other decisive sites. They talked. The rest of us didn't need to bother. We didn't need to

Enacting atherosclerosis as a process has a place in vascular surgery: in young patients blood lipoproteins are measured. The angiologist is trying to turn blood measurements into a matter of routine for all vascular patients. When should this be done? At the moment people first come to the outpatient clinic.

The angiologist: "I have proposed that, as a matter of routine, all patients who come to the outpatient clinic with vascular problems should have a blood test. In addition to their current diagnosis. All of them, not just the young ones. The professor of vascular surgery agrees, and yet it will be difficult to get it organized. It just doesn't mean enough to surgeons. So I'm now talking to the nurses, for in the outpatient clinic they do the paperwork. If they slip another form in, a good one, it may succeed."

In vascular surgery, the place for attending to bad lipoprotein levels is not before or after patients come to the outpatient clinic. It is to be located at the same time. This might be expected. In the vascular surgery outpatient clinic atherosclerosis doesn't have to do with time. It is a condition. If deviant blood levels have a place at all, it is in relation to that condition. And that is indeed where they are situated. If bad lipoproteins are enacted in vascular surgery, then this is as another *layer*. As the "underlying process." They are *underneath* the object that surgeons are able to diagnose and treat.

A vascular surgeon: "Of course we only treat the symptoms. The atherosclerosis goes on. Often we see these people back again and again. In a few months, a few years. One could get depressed about it. But then again. You cannot just send people back home once they've developed a really bad condition. You cannot let them go on until their legs drop off. Gangrene is a pretty nasty way to die. So what can you do? You say: mind your diet. Try to walk. Please stop smoking. And you keep on operating."

fight or seek a compromise. And, as disappointed radicals using Lijphart's language pointed out, those at the pillars' bottom never met and were thus kept from joining forces.

I do not refer to this book because I think you should urgently read it. Instead, I do so because you deserve to know that the image of tensions that do not turn into controversies but get "distributed" over different sites, instead, may well be a Dutch image. One that I carried with me to my field from having been brought up in

this country where so many differences get distributed—and still do, even if the traditional "pillars" are no longer so important. (See, for example, Duyvendak 1994, which describes how the gay movement in the Netherlands is divided into subgroups that do not meet but are distributed over "pillars.") Or maybe I carried this image along with me from having read the wrong political theorists—like Lijphart, who, years after 1968, not only came to state explicitly that "pillarization" is a *good* way of handling difference but also associated South

In this quote, one atherosclerosis, the process, is turned into the underlying disease. And the other, the condition, is the layer visible on the surface, the symptomatic one. This move turns operations on leg arteries into interventions that only touch the surface, not the depths. But however symptomatic in relation to the underlying disease, operations may save lives. Or at least allow people to die of a heart attack in a few years time instead of now, in several weeks, of something as nasty as gangrene.

Different diseases may clash and yet make a place for each other. The very names given to these places depend on the specificities of the disease(s). This happens with atherosclerosis. In internal medicine atherosclerosis *is* a process, but the poor condition of an encroached vessel lumen that gives complaints is a part of this process. It is to be found in a late stage. And, the other way around, in vascular surgery atherosclerosis *is* a bad condition, but deviant lipid levels in the blood may be the deep, underlying layer beneath the symptoms of the disease that are treated invasively. Instead of a global controversy or a consensus, this is another distribution of reality over different sites. Over different sites, this time, in the reality of atherosclerosis.

The Place of Blood

Hospital Z is an academic hospital. There are wings and corridors where no patient ever goes. Atherosclerosis may go there, but patients don't. In the hematology research laboratory even doctors are unusual. The research is mainly done by biologists, biochemists, and technicians. The object of investigation in the hematology laboratory at Z is blood. Or better still: the blood clotting mecha-

African *apartheid* with it—a system that he warmly recommended. No wonder his work became, in political science in the Netherlands, the object of a fierce controversy that has never closed.

Noncontradiction
In his philosophical treatise *Irreductions*, Latour eloquently sides with those sociologists who try to submit logics to sociologics. "'The strongest reason always yields to the reason of the strongest.' This is the supplement of goodness that I would like to take away. The reasoning of the strongest is simply the strongest.

'This world here below' would be very different if we were to take away this supplement, which does not exist, if we were to rob the victors of this little addition. For a start, it would no longer be a base world" (186). Elsewhere in this same text, however, Latour departs from logics in a slightly different way. Not by linking the various propositions in a controversy to "winners" and "losers," but by embedding them in matters of practice. In ways of acting, handling. In *practices* that take one in different directions: "Nothing is by itself either logical or illogical. A path always goes somewhere. All we need to know is

nism. In the hematology research laboratory atherosclerosis is enacted as a deviance that involves the blood clotting mechanism. A rupture in the intima of a thickened arterial wall triggers the blood clotting process. Platelets attach themselves to the rupture as if they needed to heal it. Thus they form more and more plaque. And debris gets attached.

The pet device of the hematology laboratory of hospital Z is the flow chamber. In the core of this chamber a slide is inserted. A fluid passes through the chamber, over the slide. It enters and leaves through plastic tubes. A pump gives the fluid a pulse-like flow. It looks like blood, this fluid, but it is made of the cells of one donor and the plasma of another. It also differs from the blood in a human body because some substances, like the fat that would immediately clog the tubes, has been washed away to make the investigation possible. But the pH is buffered at 7.4. And the temperature is human, too: the flow chamber is kept in a basin with water at 37°C. Carefully spread onto the slide inside the flow chamber is a small piece of vessel wall taken out of a "fresh" coronary artery. The question is how many blood platelets will attach to the various layers of the wall. This question is explored again and again. And each time the researcher changes another fluid variable.

The object of this research is not atherosclerosis of the leg arteries, for such localizations don't fit in with the logic of the lab. Even blood inside a body doesn't confine itself to legs, it flows everywhere. The blood investigated in the lab, moreover, was tapped from veins, not arteries. It was tapped from the veins of two different people and was carried to the lab in plastic bags. Its anatomical location is completely lost.

In the hematology lab, atherosclerosis is enacted as an interaction between blood components and the vessel wall. The experimenters manipulate the various variables of the blood that are involved in blood clotting. One at a time in an

where it goes and what kind of traffic it has to carry. Who would be so foolish as to call freeways 'logical,' roads 'illogical' and donkey tracks 'absurd'?" (179). Different roads do not contradict each other, they carry different kinds of traffic in different directions. And if "theories" are not taken to be statements about A that exclude non-A, but as diverging ways of handling reality, then a difference between them need not be a contradiction either.

Evoking practices in this way resonates with an older tradition in philosophy: that of *pragmatism*. In discussions about the *practice of* medicine, pragmatism has often been mobilized as well. It tells that the diverging disease characterizations of various medical specialisms are not meant to meet each other in the single arena of a scientific controversy. Instead, they suit different purposes. Long before Engelhardt joined Caplan in making the above-mentioned list of causes of closure, he had similarly listed diverging theories of tuberculosis. He had presented these as ever so many ways of practically handling this

endless series of experiments. This is a long way from what vascular surgeons do in their operating theaters and outpatient clinics.

The hematology professor: "I don't blame the vascular surgeons for not knowing every bio-chemical detail of the atherosclerotic process such as we describe it today. Why should they? Let them unplug pipes, they're good at that."

Surgeons do not see blood. Or they may *see* a lot of blood while they operate, but they try not to. They try to keep as much of it inside the vascular system as possible. Hematologists, on the other hand, don't see patients.

Interviewer: "Do you ever see patients with atherosclerosis yourself?" The professor of hema-tology looks surprised. I feel stupid, I should have known the answer to this question before I came to see him. "No, no, we see people with blood diseases. People with bleeding disorders, cancers. Things like that. No, we never see patients because of their atherosclerosis. So far we have nothing to offer to them."

Since hematologists have nothing to offer vascular patients they don't get to see them. Instead, they wash the blood that comes to them in bags, centrifuge it, and experiment with it. They observe platelet adhesion. They get to know the role of calcium antagonists in the clotting mechanism.

There is no controversy over whether atherosclerosis is "really" a problem of the blood clotting mechanism or, rather, one of arteries that have a stenosis or of patients hampered in their daily lives. The matter is not to be settled through dispute. It is distributed. In the day-to-day practice of hospital Z hematologists and vascular surgeons hardly ever meet. The two professors cosign grant pro-posals. Most researchers in the hematology lab usually go to the monthly re-search meeting on atherosclerosis; some of the surgeons go from time to time. But if the hematology researchers and the surgeons have different stories to tell

disease. They didn't contradict each other, but "simply" had different goals. Bacteri-ologists, Engelhardt said, fight a bug and to them tuberculosis is characterized by the microbes causing it; internists chart and treat a problem of the lungs; and since it is the task of those who work in social medicine to take care of the health of the population, they see tuberculosis as an in-fectious disease (see Engelhardt 1975, 125–41).

However, to link up the various *theories* about tuberculosis to the *tasks* of the dif-ferent physicians involved is to skip over something important: just like the medi-cal theories that help to serve them, tasks are not pregiven. They depend on and dif-fer according to the theories in question. And they may come with clashes. Either the goals of interventions themselves or the means of reaching those goals may clash. What bacteriologists, internists, and those involved in social medicine seek to accomplish may be complementary or it

me, they don't tell them to each other. Going from the department of vascular surgery to the lab is really like going from one world to another.

The junior researcher is an interesting informant because he wants to become a surgeon one day. He is scheduled for weekend and night duties along with the surgery residents. But his Ph.D. research is to do with growing media cells on artificial bypasses. His daily work is in the hematology lab. Here, after all, the cultivation of media cells is common practice. The Petri dishes, sterile work space, growth medium, stoves, and the subtleties of working with them are all ready to take up and use. But his grant givers want him to come up with results applicable in surgery within a few years, and his laboratory colleagues wonder whether his experimental setup allows him to produce insights basic enough to be publishable in any hematology journal.

"They are worlds apart. I move between them. They know nothing about each other. Nothing. It's astonishing. When my colleagues here in the lab need a bit of vessel wall, they don't know what to do. So sometimes I make a few phone calls and then I go to the operating theater when they're cutting out a piece of vessel. If I go there myself, it isn't thrown away. Instead I put it in a container and take it to the lab. Here they may find I'm not scientific enough, but my easy access to material gives me some credit. For they're all frightened of surgeons. And the barrier the other way around is at least as big. I can't explain the first thing about my research to the other surgery residents. They don't understand two words of it."

The two worlds aren't simply separated by a few floors and staircases. It's not just that in one world blood is investigated while the other operates on vessels. Their human populations are different, too. Most of the people working

may be incompatible—or it may be both (in ways that are quite similar to the dealings with atherosclerosis of the leg vessels described in the present book). By making a list that doesn't go into the character of the relations between the various medical theory/practices, Engelhardt's version of pragmatism *reproduces* the process of distribution rather than analyzes it.

There are other texts around that do not have this problem. Nicolas Fox, for instance, includes the difference between goals as a *part* of the divergence he analyses, rather than taking it as given. He has studied operations and shows that there are tensions between anesthesiologists who are concerned with their patients' overall fitness and surgeons who seek to eliminate a specific disease. A pragmatist might have tamed this difference by linking it up to the tasks of these professionals. Fox, however, digs out the tension. He doesn't do so, however, by trying to know more about its content, for instance, through going into the intertwined history of tasks, disease concepts, and technical tools. Instead, he makes a link to the sociology of professions. He shows how the tension about "fitness" versus "disease elimination" figures in the problem-ridden

in the hematology lab think medical doctors aren't scientific enough and yet feel frightened of real surgeons, whereas most vascular surgeons don't speak the hematological language. The architectural divide is duplicated by a divide between human populations.

And yet the gap between the atheroscleroses enacted in vascular surgery and hematological research is not a matter of a difference in people and their perspectives. The different worlds may be inhabited by different people, but the people do not make the difference. Even if all surgeons were to *perceive* atherosclerosis as a gradual process of plaque formation tomorrow, they wouldn't be able to *treat* it in this way. There is no working drug. And by the time there will be, it will become difficult to perceive atherosclerosis as a condition in which one has encroached vessels. By then clogged-up arteries are likely to have become rare. Most patients will never develop an atherosclerotic condition severe enough to consider invasive treatment. This will be prevented.

If there is no controversy between vascular surgeons and hematologists, this does not signal consensus but a lack of overlap between their practices. There is nothing to fight about. And yet the research that is done in the hematology lab is likely to wipe out vascular surgery some day in the future. Its aim is to find a variable that is both crucial for platelet adhesion and easy to influence. Such a drug (if used) would eradicate atherosclerosis. The popular summary of the proposal of the research project I happened to observe puts it like this:

"The influence of the different variables on the interaction between the collagens in the atherosclerotic plaque and platelets will be investigated. The project will give more insight into the basic mechanism behind the development of vascular disease. This knowledge is essential to the development of new drugs."

social relations between the two professional groups involved and are disturbing to the organization of operations (1994).

Thus, Fox's sociology is like that of many of the earlier mentioned sociologists of science who looked at conflicts between *social groups* (and their interests and social situatedness) when trying to understand controversies. There also exists sociology that relates the various "frames of reference" that physicians mobilize in their daily practice with different procedures, forms, ways of asking questions. These are not the exclusive property of some social group excluding others, but traditions, repertoires, or logics that everyone (or almost everyone) may draw on. Individuals may well get involved in more than one of them. They may shift between one frame of reference and another. This is what Nicolas Dodier describes in his study of occupational medicine. The relevant differences run right through the doctors involved. In one case, with one patient, on one day, they draw on *clinical* expertise and explanations, while in another case

The investigations are carried out by a postdoc. She's a biologist. She fills test tubes with solvents and reagents. She adds different types of collagens. She is intrigued by the lives and times of platelets, and yet well aware of the larger aims of her study.

The postdoc: "If only humans were more like rabbits. In many studies they have investigated rabbits. Lots of drugs have been found for rabbits. But they have not been successfully shifted to human models. So that's why we here in Z investigate human tissues. That's the good thing about the flow chamber. It allows us to study human materials directly."

Her prof, she says, is a doctor. He doesn't like rabbits. That's why he did so much work on the development of the flow chamber. He is also good at extrapolating whatever happens there to the atherosclerosis that develops in the arteries of humans.

The interviewer: "Do you think you'll find a drug?" The hematology professor: "Of course I do. I'd better. I put it in all my grant proposals, don't I? But yeah, sincerely, I do believe we'll find a drug. Look how many people are involved in this research. In the entire world. Mainly in the U.S., of course, but in other places as well. It's big business. The pharmaceutical industry is deeply involved. Half the current Western mortality is from atherosclerosis. And once we find something, things are likely to move fast. Vascular surgery will become superfluous. Look at ulcers. Thirty years ago a large percentage of operations were on stomach ulcers. Now all these patients are on drugs. No more operations at all."

A drug that halts the atherosclerotic process hasn't yet been found. But this is cast as a mere matter of time. The variables involved in plaque formation are numerous; the labs investigating them are numerous, too; and there is a big economic push behind that research. Sooner or later one variable or other will become manipulatable. Accessible to intervention. Vascular surgery is under threat.

they get absorbed in an *administrative* way of working. That is, here they relate to an individual with specific idiosyncrasies, normal values, styles, and problems, while there, in that other instance, they fit individuals into slots and relate them to administrative standards. The *frames of reference* Dodier differentiates between may well be *incompatible* in any single situation. And yet individual doctors tend to have no problem mobilizing one frame in one situation, and, elsewhere, a little earlier or later, the other frame (Dodier 1993).

Here, then, we have left the sociological tradition of focusing on *conflict* without shifting back into the philosophical fascination with logical *contradiction*. So where, for the time being, have we ended up? The answer is: in a place of tensions. A place were clashes may occur—or different ways of working may get spread out over different sites and situations, different build-

The resident: "So I'll be a surgeon. Yes. It will take me four more years." Interviewer: "And after that? What do you want to do, general surgery? Or vessels?" The resident: "I don't know. I like vascular surgery, it's fun. Both the craft side of it and the intellectual puzzles. But I'm warned against it. For if someone finds a drug, and preferably, that's important as well, and they're working on it, a genetic marker, a genetic marker that indicates who's prone to atherosclerosis and should take the drug, well, then it's finished. Done and over with. No more vascular surgery. Or almost."

Hematology undermines the future of vascular surgery by striving to find a drug that will make operations obsolete — especially if geneticists who are working on this in large numbers succeed in designating prospective patients. These individuals, by being pointed out to be *at risk*, are likely to take the drug and thus prevent themselves from becoming patients. So should we expect a controversy between vascular surgery and hematology in a few years time? But no. Once there is a drug, atherosclerosis will change. Claudication complaints, stenotic arteries, obstructions of the lumen, patients hampered by pain on walking: these will no longer occur. General practitioners will hardly ever refer patients with atherosclerosis to vascular surgeons any more. But they won't refer them to hematologists either: instead they'll prescribe the new drug to patients at risk and encourage these patients to take it. So once more there will be nothing to contest or fight about. The current distribution (in which surgeons treat and hematologists make promises) will fade away. It will be replaced by a quite different configuration (in which pills prevent or at least slow down the development of serious artery encroachment).

Sides and Sites

In scientific practices the shared aim is to produce knowledge that, called *universal*, can travel widely. Scientific articles try to attune their versions of the ob-

ings, rooms, times, people, questions. A place where things are what they happen to have become but could have been different — not just because they have been different in the past, but also because in fact they *are* different right now, a little further along (in another site or situation). A place quite like the one that Chantal Mouffe evokes when she warns that *difference* should be taken a lot more seriously in political theory (Mouffe 1993). Seriously —

not as a pluralism that fragments society into isolated individuals, but as a tension that comes about inevitably from the fact that, somehow, we have to *share* the world. There need not be a single victor as soon as we do not manage to smooth all our differences away into consensus. Taking difference seriously requires, or so Mouffe argues, a continuing movement between taking distance and mixing things together. Between leaving — otherness — be and re-

ject they share, and if this gives rise to tensions, they engage in controversies. Scientific alignment, however, only has a chance of success when research practices and experimental setups are made similar from one laboratory to another. Between the departments of a hospital no such practical similarity is strived after. They have, after all, different things to do; they willfully engage in different tasks. A shared, coherent ontology is not required for treatment and prevention practices. Incompatibilities between objects enacted are no obstacle to medicine's capabilities to intervene—as long as the incompatible variants of an object are separated out. This then, is what happens. The possible tensions between different variants of a disease disappear into the background when these variants are distributed over different sites. Medicine's incoherence is no flaw that requires to be mended; it does not designate a sad lack of scientificity. That the ontology enacted in medical practice is an amalgam of variants-in-tension is more likely to contribute to the rich, adaptable, and yet tenacious character of medical practice.

Distributions separate out what might otherwise clash. This chapter presented several forms of distribution. The first was the distribution of different atheroscleroses enacted over moments in a patient's itinerary: diagnosis and treatment. The "atherosclerosis" diagnosed and treated need not be the same. It doesn't necessarily pose problems if they are different. If *this* atherosclerosis is diagnosed and *that other one* treated, each variant has a site of its own. Thus, there are no competing sides to choose between or to fight for. There isn't necessarily fragmentation either, because there is flow. The object enacted does not cohere, but there is an itinerary (held together with forms, appointments,

lating to it. What is important here is the recognition that frictions are vital elements of *wholes*.

Mouffe sketches an image of the relation between different political constituents that comes close to the image of relation between different objects enacted that is sketched in the present study. And this points to a more general shift. Here, like in various recent studies, *politics* is no longer taken to be a *domain* that might or might not be separated out from the domain of *science* (see, e.g., Haraway 1997). This brings along that the relation between the two, likewise, is no longer imagined to

be a matter of the possible *invasion* from one domain into the other. Instead, what is attended to are resonances and similarities between, for instance, the *mechanics* of ways of relating. *What is it to differ?* How many styles of differing are there, how may different entities or actors both clash and show interdependence, what is the character of the "sides" involved, what kind of materials (and socials) are they made of? Such questions are as relevant when it comes to events at the level of the state as they are when it comes to a single person's fleshy life. They are as urgent where international rules are being laid down for

conversations) along which a patient may move from one site and situation to another.

The second form of distribution presented here was that of treatments over patients. Among the invasive treatments of atherosclerosis that are available in hospital Z no single one is proclaimed to be the best—be it the most effective or the least troublesome. Instead, three kinds of invasive treatment are distributed over the various patients who are taken to require invasive treatment. There are specific tools that help to achieve this distribution: indication criteria. These link up the characteristics of individual patients with either of the treatments. This, again, does not lead to fragmentation, for the various treatments come together in a central point: the place where indication criteria are set. This is the place where the object enacted and the practicalities that matter are determined interdependently.

In the third form of distribution mentioned here atherosclerosis as a *poor present condition* and as a *gradual process of deterioration* are separated out while they also acknowledge the reality of the other. This is how fragmentation is avoided here: each of the variants of atherosclerosis enacted takes the other into account. In the disease process, a poor condition is something that risks to happen at a specific moment in time: in a late stage of the process. In the disease condition, a deteriorating process is given a place in the layered body: it is the underlying reality behind the patient's condition.

And then I mentioned a fourth form of distribution. A distribution over conditions of possibility. Right now, surgical treatment makes it possible to enact

the patenting of genes as they are when it comes to the architectural and organizational design of a hospital ward. It makes little sense to call some of these sites *political* and others *science:* they all have to do with the organization of human lives and the world that comes with this, and in all of them rules, regulations, ideals, facts, frictions, frames, and tensions are paramount.

This does not imply that every site and situation is similar or that patterns of differing remain the same from one site to another. In a front-line scientific journal it makes sense to set up a difference between stated facts as a controversy, whereas in a hospital setting coordination and distributing are often more appropriate ways of handling differences. In a committee meeting of a government body it may be wise to tone down potential tension so as to reach consensus, but in a philosophical analysis bringing tensions out tends to be a virtue. This is not a call for homogenization after *politics* and *science* are no longer separate domains. Instead, it is a plea for attending to the various ways in which differences are handled in various sites and situations—and a way of wondering when and where we might do better.

atherosclerosis as a vessel encroachment that needs to be scraped away, pushed aside, or circumvented. In the hematology laboratory atherosclerosis can be enacted as a process that involves a chain of blood clotting mechanisms, but this cannot be done in the rest of the hospital—yet. A safe drug that intervenes in the blood clotting mechanism is not available: it is a mere promise. But once such a drug is on the market, the conditions of possibility will alter. Gradually, it will become more and more difficult to enact atherosclerosis as an encroached artery. Arteries that encroach are likely to become rare. At any single moment in time there is not even incoherence—let alone fragmentation. But in the course of a few years, the object atherosclerosis may have completely altered.

These, then, are four forms of distribution that keep different enactments of a single object, atherosclerosis, apart. But still the word atherosclerosis moves between the sites over which, each time, the disease is distributed. My informants do not use this word as persistently as I do. They have various local alternatives (claudication, stenosis, vascular disease, plaque formation, macrovascular complications). But "atherosclerosis" is the word they use when they want to talk to one another. The term is a coordinating mechanism operative in conjunction with the various distributions. It bridges the boundaries between the sites over which the disease is distributed. It thereby helps to prevent distribution from becoming the pluralizing of a disease into separate and unrelated objects. Distribution, instead, sets apart what also, elsewhere, a little further along, or slightly later, is linked up again. It multiplies the body and its diseases—which hang together even so.

As a Whole?

It is possible to engage in an ethnography/praxiography of *disease:* it requires that we keep the practicalities of doing disease unbracketed—in the forefront of our attention. And if we do this we learn that in different sites, different atheroscleroses are enacted. But this does not imply that the hospital explodes into idiosyncratic fragments. Instead, the singularity of objects, so often presupposed, turns out to be an accomplishment. It is the result of the work of coordination. The relative scarcity of controversy in daily practices, where so many different objects go under a single name, is likewise a remarkable achievement. It is a result of distribution. It comes about by keeping diverging objects apart if bringing them together might lead to too much friction. As long as incompatible atheroscleroses do not meet, they are in no position to confront each other.

The body multiple that ensues does not fit within an Euclidean space. In the textbook body—the single virtual body onto which various variants of atherosclerosis tend to be projected—smaller parts join together to form larger wholes. A cell is part of a tissue, tissues compose an organ, organs make a body, bodies form a population, and populations are part of an ecosystem. The precise character of the relations between the parts and what encompasses them is a matter of controversy, but however fierce the debates about this, they are based on a shared conviction: that reality is singular. Friends and foes agree that medicine should add up its dispersed findings and treat the patient as a whole. Stronger still, if it wants to do really well, medicine should take into consideration that each whole patient is part of something larger: a family (relevant for the social support it may give or the biological resemblances it may harbor), a population. The circles grow and grow. And the largest circle contains all the others.

But as soon as the practicalities of enacting reality are foregrounded, such scaling efforts collapse. A good way of beginning to show this is to attend to representational devices. In scientific journals, a picture of a chromosome is printed the same size as that of a galaxy. And take two graphs in a single book about atherosclerosis. One represents the relation between platelet adhesion on the vessel wall and the concentration of calcium in the test fluid. The second shows the number of people in the world who have died of atherosclerosis over the past ten years. How might one decide which object is *larger* that the other? The graphs may be printed in the same font, making use of similar kinds of straight and curved lines.

Once objects are taken to be a part of the practices that enact them, their sizes aren't all that easy to put in a hierarchical order. Which is larger: the "serious trouble" of the patient who is ever so sad about his immobility due to a nagging pain when walking—or the ankle/arm index of 0.7 of the next patient? Which is smaller: the atheromatous plaque taken out of a superficial femoral artery in the operating theater—or the high lipoprotein level in the blood of the patient who has been operated on? These questions simply cannot be answered. Which is larger: a reduction of 10 percent in the cholesterol intake of 100 adult males, or a successful bypass operation in one of them? Objects such as these do not have transitive relations.

This chapter is concerned with intransitivity. It shows that in practice medical ontology is not an assemblage of objects that rank from small to large. There is no framing of the patient big enough to contain all the others—and thus form

Normal and Pathological

Where the focus is on *controversies*, whether they are staged as logical contradictions or social conflicts, the image of *difference* evoked is that of opposition. In logics A excludes *non-A*. In sociology there is talk of social groups external to and in tension with another. Being in opposition, however, is not the only way to be different. There are lots of framings around of *differences* that are not necessarily *opposites*. In an analysis of medical practices one of these is of crucial importance. Medicine itself has been organized around it for a long time. This is the difference between *normal* and *pathological*.

In this book, the analysis of the way the difference between normal and pathological is made in enacting atherosclerosis is developed over various chapters. So I have hesitated about where best to insert musings about the literature on the topic. The obligation to situate them *somewhere* goes against the pervasiveness of the issues raised. But there it is: texts on paper may have a hard time dealing with linearity but they cannot avoid practicing it. Inserting notes on the normal/pathological

a "whole." Sure, in practice objects may be part of each other. When one object is enacted, another may be included in it. But this is not a matter of scale, if only because such inclusions may be reciprocal. Sometimes two objects each contain the other. In a transitive world where scale was fixed and hierarchical in character, this could never be: that A included B, while B was also inside A. But in the world of objects enacted that we live in such things happen. It is even possible that objects include one another while, simultaneously, in several ways, they are incompatible.

The As and Bs followed and unraveled in this book are variants of atherosclerosis. But it is important to remember that when atherosclerosis is enacted, lots of other subjects and objects are present on the scene. They, too, get their actuality, their shape, their thickness in and with the very activities that make some variant of atherosclerosis be. And the various objects that come together in a specific site, "doing" one another there, also depend on one another. This is what makes a praxiographic analysis so complex: that no entity can innocently stay the same throughout the story, unaltered between various sites. There are no invariable variables. There is interdependence and, where two or three modes of ordering, two or three ways of enacting a specific object meet: there is interference, too. What becomes of objects when practices interfere with one another?

Unscaling the Body

To enact a disease is also to enact norms and standards. This is because the entity afflicted by the disease deviates—from some normality. A lot might be said about normality. Here my primary concern is not with the norms that signal where deviance begins, nor with the standards by which improvement is measured. The (in)transitivity I address concerns "the entity afflicted." What is the substrate of the disease: *who* has atherosclerosis—or *what* has it? Let's see.

divide right here allows me to enrich the present chapter with a parallel discussion on modes and models of conceptualizing difference.

A lot has been written about the concept of "disease" (a variety of classical texts may be found; see Caplan, Engelhardt, and McCartney 1981). If I am to stay concise, however, and refer only sparingly to the literature, the crucial text to go into is Georges Canguilhem's *The Normal and the Pathological* ([1943] 1966). It

tells about nineteenth-century medical research in which the difference between normal and pathological was taken to be quantitative. Pathological conditions could then be studied with the aim of learning about those that were normal because they were an exaggerated or diminished form of the normal. Elsewhere, the arrow pointed in the other direction. From studying the normal function of an organ one could learn what it failed to do when it came to be pathological. But, or so Can-

If an angiographic image hangs on a light box, the fingers and the talk of those around it point to the disease of this or that specific *blood vessel*. The various arteries visible on any single image aren't in a similarly bad condition. One artery may have an 80 percent stenosis while another has a stenosis of 60 percent and a third one is not stenotic at all. A discussion about treatment will ensue. Someone may suggest that the small stenosis doesn't really require invasive treatment, but that since a catheter is being inserted in the arterial system to treat the larger one anyway, it might be worthwhile to give the smaller one a try as well. At that point the fact that the different vessels are part of a single body becomes relevant. But it doesn't stop them from being assessed individually.

Not all atheroscleroses are enacted as diseases of specific arteries. The stenoses on the light box are. But where atherosclerosis is enacted as a process of gradual vessel encroachment, precise localization in a single artery has little relevance. Gradual encroachment is a systemic disease: it concerns the *vascular system*. This implies that even if patients come to see a doctor because of hurting legs, their cardiac vessels will be suspected of being in the process of clogging up as well. And in between two visits to the outpatient clinic of vascular surgery for their leg problems, patients may have had a cerebrovascular accident or problems with the blood supply to their kidneys.

When it comes to deciding whether or not to treat invasively, atherosclerosis is not located in an artery, nor in the vascular system, but somewhere else. During such deliberations *the patient* is the "entity afflicted." The surgeons of hospital Z say so explicitly. "We do not treat vessels here, we treat patients." And they're proud of that. So what is the relation between the two, arteries and patients? Are arteries small and patients large and do the latter contain the former? The answer is: no, not in general. The state of patients does not follow from the condition of their arteries. Patients may be in a better or a worse condition than the angiogram of their arteries might lead one to expect. Clinical

guilhem argues (along with a few thinkers of the early twentieth century whose work he discusses), merely quantitative differences in function may be compatible with a good life. Being able to run faster than everyone else, for instance, is deviant, but not a disease. If we only want to term those conditions that are bothersome and *plague* a person *pathological* then we must recognize that the difference between normal and pathological is of a *qualitative* kind. It is not a matter of gradation. It is not a shift along a continuum. It is a jump, a gap, a break.

So there are already *two* images here of the difference between normal and pathological: one in which it is a difference of *degree*, of being situated here or there on a continuum; the other in which it is a difference in *kind*, which implies that a crucial jump is made, a gap is crossed, when one goes from one side of the bound-

atherosclerosis, medicine's assessment of "the patient's" disease, is not based on what can be seen on an angiographic image of the arteries; it is not a larger circle around this, but rather it is a reality of its own.

In daily hospital practice, arteries and patients do not have a transitive relation. Instead, they are distributed over different sites. The patient speaks in the outpatient clinic while the artery is enacted as a deviant entity in the radiology department. Or: first the patient speaks, later on the arteries are treated. So the reality of deviant arteries is not situated *inside* but *alongside* that of sick patients. This implies that interventions in vessels aren't a matter of "reducing" patients to vessels. Something more complex is going on. Let's go to the operation theater. A patient is anesthetized. The surgeons have cut the skin, the fasciae, and the muscles in order to have access to the arteries. They cut in and sew up arteries. That is what they concentrate on. But this isn't a reductionist privileging of a part that impedes a wide view on the whole. What happens during an operation is that surgeons and staff concentrate on one object, the arteries, rather than on the other, the patient. But then sometimes they make abrupt switches between the two.

We're in the operating theater. I've been away for half an hour somewhere along the way to have a coffee: I was worn out from all the flesh and blood. From the ripping apart of bodily tissues, so careful but so ruthless. From the enormous amounts of fat cut through. From the search, with sensitive hands as well as watchful eyes, for the target arteries. From the smell of the small side vessels when they are burned in order to close them off. From the cutting. But now it's almost done. The resident is making the final sutures. The fasciae. The skin. While looking at his working hands, he continues the conversation he's having with

ary to the other. In Canguilhem's work the importance of stressing this difference is to defend *the clinic* against *the lab*. Laboratory measurements, imaging technologies, and all the rest of it allow only the recognition of what is uncommon, deviant. But Canguilhem argues that whether or not the conditions thus detected plague patients only appears in the clinic where patients relate their own, singular story. The normativity that matters is clinical: laboratories can establish facts, not norms. Historically, Canguilhem adds, the clinic came first as well. Laboratories would never have been built if it were not for the existence of consulting rooms where patients come to see doctors, tell them about their complaints, and ask for help.

In Canguilhem's work, attentiveness to the clinic is argued for in a normative way: the clinic *should not* be overruled by the lab, but take the lead instead. Here, I have taken up this concern in an empirical way. Walking around in the hospital, I have asked: how do clinical diagnosis and laboratory diagnosis relate in the case of atherosclerosis of the leg vessels?

the junior surgeon. This seems to be gossip about a nurse, a mutual friend, whoever: "She's a neat person, isn't she, I like her, you can laugh with her." It takes me a few seconds to realize he's talking about the patient.

Switches between attending to arteries and attending to a patient aren't made at any random moment in an operation. When things are difficult and everybody's concentration is required, there are no shifts. But in the calm moment when the last sutures are being made, there are. "Will you phone his wife?" one surgeon may ask another when they have reached that point in an operation—thereby almost effortlessly turning the physical being on the operation table into a social one. Into someone who happens to have a wife who is likely to care about him.

Such switches do not turn blood vessels into a small part of patients. Instead, they are, indeed, *switches*. They turn the operation from an intervention in one or more arteries into an intervention in one or more lives. They do not zoom out from the details under the skin to the patient as a whole, but move the camera sideward and focus it on another object. One scene, that in which the arteries are central stage, is left and the action moves onto another one, in which the leading parts are played by people. The modes of reasoning and the necessary skills switch accordingly. In the operating theater surgeons must have steady, dexterous hands. In the outpatient clinic they need to be respectful and attentive—or so they teach their students.

Resident: "It's what I like about surgery. You have to do such different things. Talking with people, I like that, I wouldn't want to miss that. I wouldn't want to be a radiologist or

I have not left the hospital or, rather, the medical network of which it forms such an important part. Others have done so. They have taken up Canguilhem's plea to put the patient's individual suffering first as a point of contrast for the analysis of various ways in which *pathological* was/is used as a judgment that marks individuals in a negative way. This judgment differentiates some people, the deviants, from others, who are thereby taken to be standard. There is a lot of literature about the way in which this worked in the late nineteenth century. This was a period when women were marked as sickly deviants in contrast to the standard man; when blacks acquired the status of unfit and invalid human exemplars falling below the standard set by whites; when the category of the homosexual was invented to encompass people who were marked as developmental accidents failing to meet the maturity of heterosexuality. These various polarities, all feeding on the difference between the *normal* and the *pathological*, were linked together. They informed and colored one another (see, e.g., Gilman 1985; Stepan 1987; Showalter 1985). Writing about these tropes forms part of the attempt to escape from their continued

an anesthesiologist, I like to have contact with patients. With people. But I also like the precision, the technicalities of an operation. The craft side of it."

Surgeons may be unkind or make mistakes when operating. They may excel or fail when it comes to talking or the craft side of their work. But all surgeons, good or bad, have gotten used to making switches between repertoires. For the passing ethnographer this is more difficult. Like many outsiders I had trouble facing the reality of the skinless flesh that becomes visible in the operating theater. I had to put in effort to collide in enacting a reality I wasn't used to, and such a bloody one at that. No wonder it took me a while to realize that the resident was talking about the neatness and the sense of humor of the very patient he was sewing up.

Surgeons usually know the patient on whom they are operating from the consulting room and ward rounds. But the art of switching from organ to patient doesn't depend on this, for even pathologists who never knew the patient they are dissecting have this capacity. The reality of the patient—or at least some version of it—is sustained even after treatment has failed and only a corpse remains.

A corpse lies on a high steel table that has small holes in it through which fluids can drip away. A resident dissects. She's assisted by a technician who, when necessary, handles the saw and sucks up the blood with a small machine. When the resident takes her scissors to cut the aorta, she warns me: "Listen! Yeah! Do you hear that? There's your atherosclerosis." I hear it. A cracking sound. Calcification. I want to make a note: during a dissection atherosclerosis is an audible calcification of the vessel wall. Before I have been able to write it down, another pathologist passes by. How are things going? The resident says she's doing fine. But could her colleague please have a look at the patient's eye? There's a strange blue

influence in our present-day conceptual apparatus.

A crucial step in the attempts to *analyze out* the imposition of the pathological on different kinds of othernesses is the work of Michel Foucault that I have mentioned a few times earlier (especially Foucault 1973). He showed that the polar distinction between "normal" and "pathological," however pervasive it was in the nineteenth century, is not all that old. It is indeed no older than that nineteenth century. Before that time *disease* was not

taken to be a *condition of* the body, contrasting with that other condition, health. There were *diseases* and they could come to *inhabit* a body. The crucial difference to attend to was not that between one body (normal) and another (pathological), but between one disease and another. Making differences, then, was a matter of making classifications. Classifications in which the diseases were listed in nosological tables, just like Linnaeus listed plants. Diseases were like species and doctors had to try to recognize them through the transpar-

tinge around it. The colleague nods and removes the cloth that covers the head. He inspects the eye and then puts the cloth back again.

In the dissection room the technical tool used to switch repertoires is a cotton cloth, fifty centimeters wide and fifty centimeters long, of a vague off-white shade. That's all. A corpse with cracking arteries is not smaller than the patient-as-whole. It is not an ingredient out of which people are made (all you have to do is to add some life to it, and there you go). There are, instead, two creatures. One is having its insides taken out and its organs are being cut into slices. The other is being accorded human dignity and treated with respect. She's even being spared the sight of her own dissection. Her persona, residing in her face, is kept out of the pathologist's permanent field of vision.

The pathology technician is happy to tell me about his hidden work. "So after the dissection, I have to sew them up again. I wash the blood from the skin, that's tough. It's good to be careful with blood, it's sticky, hard to clean away. And I fill up the belly and the thorax and dress the body and try to do it all in such a way that the family won't notice. They have to say goodbye to their mother, their sister, their wife, whatever. That's difficult enough. They shouldn't be worried too much with what we have been doing to the body."

Maybe the cloth is not a symbolic device, but a practical one. Instead of obliging the doctors to respect the patient, it protects the head from getting stained by the blood that may splash at the rougher moments of a dissection. Or perhaps it does both. Whichever way, for the technician turning a dissected corpse back into a person fit to be buried involves a lot more work than just lifting a piece of cloth. He has to fill the cavities out of which the organs are taken away, sew up the skin, clean it, dress the naked body. Caring for a cold corpse that doesn't smell good is hard—but it is important. If the technician does it well, he spares the family the task (a lot harder for them than for well-trained profes-

ent, but sometimes misleading, body. The body's transparency ended in the early nineteenth century when disease came to be treated as a pathological state of the tissues. A state that was opaque in the living and could only be unveiled by opening up some bodies.

Foucault's way of presenting this history was intended to rob the differentiation between normal and pathological of its supposedly natural character. This dif-ferentiation only developed with *the clinic:* a word, this time, that designates a specific way of organizing hospitals and the education of young doctors. It does not speak from the body all by itself but is a specific practically and materially organized way of making the body speak, which means that one day it might lose its authority. Maybe this has happened already—or is in the process of happening. Foucault contributes to this erosion by showing the tempo-

sionals) of shifting repertoires. He keeps aortas that crack when they are cut in with scissors out of the social life of the deceased person.

A Tension and a Loop

In some repertoires atherosclerosis is a disease of arteries; in others, patients suffer from it. And there is yet a third entity that may be plagued by this disease: populations. Let's take a closer look at the relation between the atheroscleroses of individuals and that of populations. It is, as might be expected, quite complex. In an individual patient's life atherosclerosis is usually only one of a series of problems. Atherosclerosis, however bad it is, is not the only reality that patients live with.

Internist: "You should be careful about just talking about 'atherosclerosis' because you hap-pen to be interested in that. Most of these people have lots of things. They may have diabetes, I gather you've come across that. People who are in a really bad condition, especially those who've had an amputation, tend to have diabetes. Or a lipoprotein disorder. And then some-body may have bad lungs, asthma, whatever. It's unrelated, but it's what they have to live with. And another patient has great problems because he's just had to quit his job. And a third—what have you—a third has neurological problems, or a nagging huge wart on their feet that makes it impossible to keep on walking."

The atherosclerosis that plagues a patient is just one of the many elements of that patient's life. There are others. Other diseases. But also other kinds of phenomena, like work, or grandchildren, or gardens. In the patient's file this life is not summed up: only the so-called medical problems are listed, one after the other.

rality of the division between normal and pathological—and its practical base. Thus, he suggests how all those classified as *ab-normals* might escape from that category, not in order to end up in the other one, that of the *normals*, but in order to end up else-where, without either of these identities.

And how does that literature figure in the background of the present book? One answer is that it makes one wonder if what happens at the present time should be caught (as I did above) in the ten-sion between the clinic and the lab—where the clinic is taken to be a clinical way to establish *ab*-normality. What is currently being established in the clinic might well be something slightly, but crucially, differ-ent. The medical question par excellence is no longer the question Foucault pointed out as such: "Where does it hurt?" Instead, it has become this other one: "What is your problem?" This is a question about whether you, the patient, are still able to live a good life, or whether you have *a prob-lem* with that. The problems one is faced with are not conditions of the body. They pertain to one's body, but they are situated elsewhere: in one's life. With this comes

*"Mrs. Linder. Female. Date of birth: 15.03.1937. Patient is admitted with severe claudica-
tion of her right leg. Some rest pain. Poor skin. Ankle/arm index 0.8. Duplex: stenosis in
right popliteal artery. Other problems: overweight, repeated hernias, glaucoma."*

Mrs. Linder has various problems. Those the resident asked about and takes
to be relevant are written down in her file when she's admitted in order to have
an angiogram. Thus, in files, problems come together. The admission itself also
enters another administrative system: that of the hospital. The hospital adminis-
tration counts admissions. It needs to do so in order to be able to send invoices to
the insurance companies. And since it counts admissions anyway, it is also able
to feed these to centers for the study of *epidemiology*. There all admissions for
arterial disease of the leg vessels are drawn together to enact atherosclerosis as
something that afflicts a population, such as the population of the Netherlands,
the country where hospital Z is situated.

*In the Netherlands in 1992, 170 men of every 100,000 inhabitants and 70 women of
every 100,000 inhabitants were admitted to a hospital for peripheral arterial disease. (The
source of these numbers is a report of the Dutch Heart Foundation on women and heart
and arterial diseases: Vrouwen en Hart- en Vaatziekten, Nederlandse Hartstichting, Den
Haag, 1994.)*

The admission of Mrs. Linder must be somewhere among those numbers.
It's included in the calculation. Mrs. Linder's other problems are erased along
the way. So however big the difference between Mrs. Linder's stay in the hospital and that of, say, Mrs. Bonder, they both end up among the seventy women
admitted out of every 100,000 inhabitants of the Netherlands in 1992. Thus the

---

another shift: that of the subject of normativity. The professional, or professional
knowledge, is no longer the sure authority
able to differentiate between what is and
isn't a problem in a person's life. *Is this a
problem* for you, Mrs. Sangers? This is the
new trope: that patients are being elicited
to articulate norms about and for themselves.

This shift has been described before,
in the literature. (Crucial citations include
Armstrong 1983; Arney and Bergen 1984;
and, if you like reading Dutch, see also Mol
and Van Lieshout 1989.) But the pervasive

emphasis of the present book is different
because it does not tell a history in which
some things happened in the past and
others are happening in the present. No
shift here. In this book the various patterns
of differentiating mentioned are shown to
all interfere with one another. Differentiating between normal and pathological has
not ended. It *coexists* with asking a patient
how she is doing and what she does or
doesn't experience as a problem. Those
complex interferences deserve further attention.

population known by epidemiology is not "bigger" than the individual patients that compose it. Epidemiologists cut slices, not through organs, but through lives. They add up admissions, leaving out anything else that might be relevant in the lives of the individuals involved. And not just in their lives: in their diseases, too.

Epidemiology tables do not only erase people's various other problems, they also incorporate only one out of various possible ways of enacting atherosclerosis. A table of "hospital admissions" takes patients who are admitted into account. Patients who have pain on walking or a severe pressure drop but who are not admitted are not counted. Since duplex requires no hospital admission and angiography does, patients who have a deviant duplex are not included in the above numbers either, but patients in whom an angiography is made are. This shows that an epidemiological table that counts "hospital admissions" depends on the current state of diagnostic technology. Every single angiographic procedure that is replaced by a duplex scanning implies one less admission—and so one less case of the disease counted in this specific epidemiological table.

Such complexities are a matter of constant concern to epidemiologists: which of their numbers tells what? What do "indicators" indicate? For instance, which population is hit harder by atherosclerosis: that of men or women? At the time of my fieldwork this question was, in the Netherlands, turned into an issue for debate. It isn't easy. If epidemiologists count atherosclerosis of the leg vessels (by counting the number of those who die from this disease in every 100,000 inhabitants), then men suffer more than women. Cardiac vessels also clog up more in men. But these numbers may well be biased because often only male and female populations under the age of seventy are taken into account—women die from atherosclerosis older. And cerebrovascular accidents also have a comparatively high incidence in women. Thus, the picture of who suffers from atherosclerosis most alters if all vessels and all ages are taken into account and, again, death is taken as the indicator.

Self and Other
One of the piles of literature I watched myself assembling over the years I studied *differences* is composed of texts addressing the way boundaries are made between what ends up being the *self* and what is differentiated from this as being *other*. Making such a differentiation has been crucial to various disciplines in the twentieth century. (For an overview of the biological and biomedical among these, see Barreau et al. 1986.) Prominent among these is a branch of science that explores how an *organism* recognizes what is part of itself and what is alien to it: immunology. That the question of where an organism begins and ends isn't obvious has already been addressed in one of the older but well-remembered studies of medical knowledge, that of Ludwig Fleck

"For women as well as men heart and vascular diseases are the number one cause of death, in the Netherlands and almost all other Western countries. In 1992 26.725 women and 25.478 men [in the Netherlands] died from the consequences of heart and vascular diseases. This made the contribution of heart and vascular diseases to the total mortality 42% for women and 38% for men." (The source of these numbers is the same report of the Dutch Heart Foundation, on women and heart and arterial diseases: Vrouwen en Hart- en Vaat- ziekten, Nederlandse Hartstichting, Den Haag, 1994.)

What epidemiology makes of the diseases of populations depends on the individual enactments of these diseases it takes into account. We might expect as much. For if there are different enactments of atherosclerosis that come with different events, and if statistical tables are an aggregation of events, then it isn't all that surprising that the tables of the disease vary together with the events counted. So, the atherosclerosis of a population depends on the variant of the individual's atherosclerosis that it *includes*. But the more surprising thing is this: it also happens the other way around. Yes, the other way around. The events that happen to individuals depend on and vary with "the population" that they, in their turn, *include*. The way individual disease is enacted depends on epidemiology.

General practitioner: "They keep on teaching you. That doing a test for a disease is not very helpful in a population where that disease is rare. Take this patient, a young man, he's heard of claudication—from his father, uncle, neighbor. And he's convinced that that's what he's got. But I talk with him and think: oh, no, this is something else. Then, what happens? I can do pressure measurements, we've bought a small Doppler and I'm able to use it. But, say he's thirty-five and his legs hurt during the night, when he's in bed. In a case like that my blood pressure measurement has a far higher chance of giving a false positive result than of finding real disease. So what do I do? I don't measure ankle pressures. It's better not to."

([1935] 1980). If an organism is a viable whole, Fleck remarks, then one may wonder whether the bacteria living peacefully in human intestines are part of the larger human organism, or not. But if the self of bacteria may fuse with that of the humans they happen to live in, then the entire ecosystem of which humans form a part may well be designated as a viable whole —an organism—in its turn. Opening up the boundaries around what seemed a self-evident whole, the organism, goes together in Fleck's text with opening up the boundaries of *science*. That, too, he shows, is not an impermeably closed off self. Its boundaries leak. Ideas flow in from elsewhere, to come together in scientific disciplines, and in the process they are gradually adapted. One doesn't need a closed boundary to defend and be oneself. Or to acquire good medical knowledge.

Meanwhile, however, at the very time that Fleck was writing, the idea gained strength that organisms only stay healthy

A diagnostic test is more likely to fit with the results of other tests if it is used in a population that contains a lot of disease. In a population with only little disease, the test is less trustworthy. Thus, the technicalities of diagnosing individuals depend on the severity of the disease in the population. And individual cases include the epidemiological aggregate in other ways as well. For instance, the very *criteria* used to judge test outcomes often derive from population studies.

An internist: "The average cholesterol in the population used to be the norm. If, say, 6.3 mmol/L was the average for males in a certain population, labs used to print this on their forms as the normal value. Then if someone had a cholesterol of 6.2 everybody was happy with it. Because it was below the norm. Treatment was only considered in men who were above the norm, say if someone had a cholesterol of 6.7 mmol/L. In some places it's still like that. There are general practitioners who even put off talking to their patients about cholesterol until it is 6.9, or 7.3. They feel it shouldn't be pressed too much."

This is a common procedure. The population average becomes the individual's target. This implies that every individual case, every encounter in the clinic, depends on the population average epidemiologically established. It is this average that determines whether or not someone is diagnosed as deviant and will be treated. But the internist just quoted doesn't agree with this use of population averages. He advocates another method for setting norms.

The same internist: "But it is a bad idea to take the average as a norm. For if you take an average Western male population, well, all these men have cholesterols that are too high. Or almost. In places where people live more healthy lives, puff, the numbers drop. So now targets are set differently. That is by looking into the relation between cholesterol levels and

if they manage to keep out or otherwise defend themselves against all invaders. Contamination was to be avoided. Wash your hands and do not kiss or spit if you happen to have tuberculosis. This idea not only led to large programs of hygiene that were meant to keep out *all* people healthy. It was also applied to *the* people, the population, as if this, like the individual, was an organism in its own right. The population—or the race, the words could be used interchangeably—should not be stained by foreign blood. In analogy with corporeal hygiene, racial hygiene became a meaningful concept. Fleck, Polish Jew, was to be con-

fronted very directly with racial hygiene, in its blunt anti-Jewish Nazi form. Around the world it took on a variety of other shapes, up to and including the South African laws against so-called mixed marriages that were still enforced in the 1980s. (For the conceptual sides of all this, see, e.g., Stocking 1968.)

On both levels, that of the individual and that of the population, the idea that there is, or should be, a single, stable boundary between *self* and *other* has been undermined over the past decades. These two movements come together in an interesting book that I picked up in the bookstore

the development of atherosclerosis. Risk of death from atherosclerosis. We've tried to find the cholesterol level that correlates with a really low risk. Right now we think it is below 5.5 mmol/L. Maybe it is even lower."

In his mistrust of the average cholesterol level in the population, this informant doesn't propose that we should deal with individuals one by one. He suggests, instead, the inclusion of another kind of population study. He doesn't want the overall male population to set the norm for men, but only that part of it that doesn't die from atherosclerosis. Thus, a cholesterol level that is so low that atherosclerosis never gets severe is proposed as the one that is good and healthy. If this cholesterol level would be taken as a norm, different assessments of individuals would result and different treatments would be proposed. The way individuals are diagnosed and treated depends on the reality of the population that is included when it comes to setting norms.

Instead of a transitive relation in which the small individual is contained in the larger population, what we find here is *mutual inclusion*. A population is an aggregate of events that happen to individuals. But the events that happen to individuals are in their turn informed by the framing of the population they belong to. The so-called whole is a part of its individual elements no less than the individual elements form part of the whole. Occasionally this may lead to circularities.

The professor of epidemiology of the university linked up with hospital Z is intrigued by the similarities and differences between the atheroscleroses of women and men. What, or so she wonders, does the gap between the various epidemiological statistics about severity mean? That women indeed suffer less from atherosclerosis in their legs than men; that they have

because of its title and its table of contents, even though its topic seemed distant from atherosclerosis of the leg vessels. This was because it seemed to be important for understanding *difference*. And so it proved to be. In *Logiques Métisses* Jean-Loup Anselle gives a historical analysis of the constitution of identity and especially ethnic identity in West Africa. He tells that in the time before colonial rule it used to be possible there to change one's name as well as the "ethnic" group to which one belonged. "So the notion 'person' or 'identity' is not a part of an immutable metaphysics. . . . The notion 'person' is persistently negotiated and contested among groups that partake in the same political unity as well as among neighboring political unities. With the appearance of the civil state and the written registrations of identities that resulted from that, it has gained a greater stability. It became much harder to change identities or even the orthography of one's name" (Anselle 1990, 203).

The French wanted individual people to state a name and an ethnic identity that they could then write down in their files. Their political system was one of fixed

it later in life, when they're (deemed to be) too old to treat; that they present their complaints differently; or that doctors for whatever reason underestimate the prevalence of this disease in women? In our interview she explains the difficulties epidemiologists confront when trying to address such questions properly.

"You see, it's always a problem, it's a problem of making data. Take mortality statistics. They cover the entire population, which makes them popular among my colleagues. But they are as fragile as anything else. Imagine what happens. Epidemiology says men have more heart attacks than women. So fine, all clinicians are taught that this is so. Then there's a sudden death. Somebody old. Good, the general practitioner goes to the patient's house, it's late at night, he wants to get home. But there are forms to fill in: cause of death. Suppose the patient is a man. The general practitioner does a quick physical, asks a few questions of the family. And since a 'heart attack' is a statistically probable cause of death and the story sounds more or less like a 'heart attack,' he writes down 'heart attack.' Nobody is going to be surprised, nobody will look into it. But if it's a woman, he may well take a heart attack to be less likely. Epidemiology, after all, has taught him that it is less likely. So he looks a little more closely, asks a few additional questions. And then the 'cause of death' may become something else. I don't know what: it doesn't matter. A cerebrovascular accident, an asthma attack, or even food poisoning. Whatever. So there's the loop. Both forms about cause of death are fed into the computer and there we go: men die of heart attacks more often than women!"

The diseases of population and patient are interdependent. The coconstitution is mutual. The elements create an aggregate and the aggregate informs the elements, which is how the atherosclerosis of the individual and that of the population in which it is included may get trapped in a circularity. They may loop. And spiral.

identities. Before modern bureaucracy was established in West Africa a person could cross boundaries, become a slightly different self, one with another ethnic identity. But not afterward. Thus, French registers helped to *constitute,* in practice, the strict boundaries around ethnic groups that they could later designate as culturally given.

The separation between *self* and *other,* then, is a separation that exists because it has been *made* to exist. Dorinne Kondo, studying *the self* in a Japanese workplace, tries to not go along with this produc-

tion, but to lay it bare (1990). Aligning with Butler, Kondo writes: "The conventional trope opposes 'the self' as bounded essence, filled with 'real feelings' and identity, to a 'world' or to a 'society' which is spatially and ontologically distinct from the self. Indeed, the academic division of labor recapitulates this distinction in its separation of the disciplines, distinguishing 'psychology' from 'sociology'" (33–34). The very practice of academic research, by either focusing on psychological or sociological phenomena, reaffirms that

Frictions Incorporated

The atherosclerosis of a population and that of individuals mutually include each other. But there are also frictions between them. There are frictions between what it implies to take as a target for health care the improvement of the conditions of individual patients or rather the improvement of the health of populations. These frictions, however, do not impede the mutual inclusion of individual-oriented and population-oriented health care. To begin unraveling this, it is important to first note that "improvement" is no straightforward matter. Take the case of Mrs. Zoka. She's been admitted to the department of vascular surgery lots of times by now. A few days ago she had a PTA. Has she improved or not? In order to answer this question we need a scale for making comparisons. A scale along which "better" and "worse" make sense. There are various possible scales or criteria for comparison.

Mrs. Zoka: "It's warm again, my leg. I can feel it's warm again, it's such a good feeling. Yeah, it still hurts, from the sutures. But I've started to walk even so and it was amazing. I was close to giving up, for I've had so many treatments, but I'm really glad they tried this."

When Mrs. Zoka compares her own condition now with how she was last week, she signals improvement. But when the axis of comparison is changed, this assessment does not need to hold. Look at Mrs. Zoka's patient history.

My informant, a surgery resident, looks up Mrs. Zoka's file for me. It's thick but it isn't the thickest one in the rack of files. On the cover we find its essentials. Name: Zoka. Sex: female. Date of birth: 07.05.28. At his desk the resident opens it. There's Mrs. Zoka's history. It's marked down that she had an appendectomy in 1948. A hysterectomy in 1967. In 1975 she was discovered to have diabetes. She was put on insulin. Moreover, since 1982 she's been being treated for high blood pressure. Her first vascular operation was in 1986.

selves and others are distinct and separate objects. Kondo suggests we might work our way out of restating and therefore reinforcing this boundary "by asking how selves *in the plural* are constructed variously in various situations, how these constructions can be complicated and enlivened by multiplicity and ambiguity, and how they shape, and are shaped by, relations of power" (43). Maybe there are many selves, implicated in many relations. They do not stand in opposition to a single outside world to which they both belong and are strangers. They are, instead, implicated in different practices.

Many selves and various others. This sounds rather similar to what you have been reading about here: different entities called atherosclerosis, generating individuals, that may be added up in various ways to form different atheroscleroses that create populations. Kondo's work on selves in a Japanese workplace, however far from hospital Z, is similar to the present

*A femoropopliteal bypass in her right leg. Here's a second one, in 1988: another femoro-
popliteal bypass, this time in the left leg. In 1989 she had a small heart attack. In 1990 a
stenosis in the vessels of the right leg was widened with a PTA. In 1991 there was a new operation. This time Mrs. Zoka's surgeons inserted a femorotibial bypass in her left leg because
the earlier bypass was clogged up. At the very moment we are inspecting her file, Mrs. Zoka
is in the hospital again. It's 1992. Last week one of her bypasses was opened up with a PTA.*

Mrs. Zoka's history is full of problems. Some are past if not forgotten; others
stay; yet others grow. The severe pain and persistent blood loss that formed the
indication of her hysterectomy have disappeared. The diabetes will go on forever. And Mrs. Zoka's arteries grow worse and worse. One intervention in her
legs is followed by another. A first heart attack indicates that her coronary vessels are involved as well. And no doctor would be surprised if Mrs. Zoka were
to have a second heart attack soon. Or a cerebrovascular accident.

So she's better than last week. But she's worse than five years ago, for more
and more vessels are gradually clogging up. And here is a third evaluation.

*Mrs. Zoka's general practitioner thinks she's doing better than five years ago. At that time
he noted down visits for coughs in his files. Sleeplessness. Trouble with getting her blood
sugar under control and her diabetes medication adjusted. Trouble with high blood pressure. Worries about her sick husband whom she took care of for years and years. By the end
he became demented and was often aggressive. He prevented her from going anywhere: a
visit to the doctor was the only way she got out of the house. Mrs. Zoka's general practitioner
thinks she's doing better since her husband died. Even if she says that she misses him, she
expresses fewer complaints these days. Now it's just her arteries that cause complaints.*

study in this respect: it also argues that
multiplication may, among other things,
lead one out of this binary opposition that
is also an entanglement: the *self* versus
the *other*.

Boundaries

Relating to the literature allows one to
link up a study on atherosclerosis in
a Dutch hospital with historical anthropology about a large region in West Africa
or an ethnography done in a neighborhood
of large Japanese city. Thus *boundaries*
are crossed, boundaries between object
domains. There is nothing extraordinary
about this: it is a conventional academic
way of crossing boundaries, one in which
theory (that is, the concepts mobilized in
making sense of the world) are made to
travel between fields of study. In that sense
theory resembles the *networks* that cross
boundaries as well. How to put this? The
dominant Western ways of framing what
belongs and what does not, what is similar
in kind and what of a different category, are
regional in character. They lump together
what is of a similar kind and imagine, or
create, a boundary around it. What is different, then, is also elsewhere. This is ex-
emplified in the process of the formation

A patient history gets its shape in a file: different files shape different histories. In the general practitioner's file the "vascular system" is one reason among many that a patient might have for a consultation. So even if her blood vessels keep on clogging up and her diabetes is there to stay, Mrs. Zoka's general practitioner may see an improvement, an improvement that is related to the sum of her other problems: for these are decreasing.

Improvement due to the treatment. Deterioration despite many treatments. Improvement that has nothing much to do with the treatment of her vascular disease. Mrs. Zoka's history doesn't unequivocally point in one direction. It is hard to evaluate it overall. It allows for different evaluations. So on an individual level "improvement" isn't straightforward. And obviously the aggregated assessment of medical interventions will vary depending on the scale and criteria used on an individual level. This, however, does not imply that a positive evaluation of the care given to an individual also implies that the health of the population has improved. Look at Mrs. Zoka's PTA of last week. When asked whether it improved Mrs. Zoka's condition, the vascular surgeons and Mrs. Zoka herself answer this question with "yes." And they would again want to do this PTA if they were confronted with the decision for the second time. Even if it doesn't lower the risk that, in one way or another, Mrs. Zoka will fairly soon die from atherosclerosis, it makes a difference to her now. But this PTA that worked so well for Mrs. Zoka doesn't improve the overall health of the Dutch population. There is a tension here.

"A one-to-one approach is of value to the patient, his family and friends, but does little to alter the distribution of disease in the population" (Syme SL, Guralinik JM: Epidemiology

of modern states in Europe. In the process of that formation, nation states came to coincide with geographically bounded territories. Their limits were marked with boundaries, fence posts, and custom officers. (The classic text to mention here is Poulanzas 1978.) In social theory, this separation work was for a long time implicitly accepted. When *society* was mentioned, it was held to reside within some state's boundaries.

Over the past decades, this restriction has been broken down in several ways. The *boundary* has become a contested

issue. First, *crossing* boundaries became a widely shared ideal. The line dividing what is similar from what is different was to be questioned. Donna Haraway is one of the spokespersons of this transgressive zeal. The boundary or the border that she puts center stage here is one I haven't gone into, even though there is much interesting work to relate to. This is the boundary between body, organism, and machine. "In the traditions of 'Western' science and politics—the tradition of racist, male-dominant capitalism; the tradition of progress; the tradition of repro-

and health policy: Coronary disease. In Levine S, Lilienfeld A, eds., Epidemiology and health policy. *New York, Tavistock, 1987, p. 106).*

If "the distribution of disease in the population" is assessed in terms of, for instance, overall mortality, Mrs. Zoka's PTA doesn't have a positive effect. Other efforts might. For the same money, or so public health promoters keep pointing out, the overall mortality from atherosclerosis might be considerably reduced. If only this money was spent on other interventions. If only the effort of all those professionals now bowing over legs and inserting catheters in arteries was directed differently.

"Concern for the welfare of individuals may be good for these particular people, but concern for the health of the public as a whole points us in a different direction. We need to consider the implications of a situation in which a small risk involves a large number of people, who in the high-risk strategy would be categorized as normal. The result for the population may be a lower number of cases, even though no one was at a conspicuous risk" (Rose G: The strategy of preventive medicine. *Oxford, Oxford Medical Publications, 1992, p. 14).*

What to do: one may treat people with disease; prevent the deterioration of people at risk; or, the best way to lower overall mortality, try to improve the health of a large group of almost-normal people. These interventions do not clash in principle: in principle they may all be done, simultaneously, one next to the other. But in practice health care money can only be spent once, and all the effort of the necessarily limited amount of health care professionals that goes into individual treatment is lost for interventions in the population.

In order to reduce mortality from atherosclerosis in the population, everybody should stop smoking, do more exercise, and go on a better diet. Such answers come out of population studies. Should population studies then come to

duction of the self from the reflections of the other—the relation between organism and machine has been a border war. The stakes in the border war have been the territories of production, reproduction and imagination. This text is an argument for *pleasure* in the confusion of boundaries and for *responsibility* in their construction" (1991, 150). Arguing for pleasure as well as responsibility, Haraway came up with a boundary-confusing image: that of the *cyborg*. Cyborgs live in two countries: that of the machine and that of the organism. "By the late twentieth century, our time, a mythic time, we are all chimeras, theorized and fabricated hybrids of machine and organism; in short we are cyborgs" (150).

The cyborg has become a much used word and image. Open up any culturally versed journal and you are likely to find it everywhere. Another term that plays similarly (but also differently) on the *blurring* of boundaries is that of the *boundary ob-*

reign in medicine and direct the efforts? There is a problem with this. The problem is that improvements of the health of populations do not necessarily have serious advantages for individuals—whether patient or nonpatient. Take cholesterol. For a long time Western populations have been encouraged to adapt their diets so as to lower their intake of cholesterol. More recently, moreover, drugs for lowering blood cholesterol levels have been developed as well. But is a reduction of cholesterol levels in the population beneficial for its individual members?

One of the professors of internal medicine of hospital Z argues that we should all lower our cholesterol levels. He says this in an interview: it's in the newspaper. He attacks Dutch general practitioners. They are responsible, he says, for thousands of deaths every year because in their protocols they have set the target cholesterol level far too high. The journalist got interested. He has been to see a general practitioner on the committee who designed the protocol. The general practitioner explains what reducing the risk of mortality in a population means to the individuals concerned. He quotes a recent study. It investigated the effects of lowering high cholesterol levels in middle-aged males. Out of the 3,302 patients who took medication, only 143 had a heart attacks. In the 3,293 controls there were more heart attacks: 204. This is a reduction of 31 percent; 31 percent fewer heart attacks in a population is a good result.

But what does this "good result" mean for the individuals who participated in the study? First, that 3,159 of them have taken medication and gone to their doctor for checkups without any personal benefit. Second, that for every individual on medication the chance of not dying of a heart attack within the five years of the study only increased from 93.5 to 95.4 percent. That doesn't sound like a worthwhile improvement for an individual. And then there's a third step. The irony in this particular study was that the overall likelihood of living through the five-year period of the study was 95.9 percent in the group on cholesterol medi-

ject. It comes out of a single, modest article that because of this wonderful word has been quoted all over the place (Star and Giesemer 1989). The concept of the boundary object grows out of the idea that there are different social worlds. These different social worlds each have their own codes, habits, instruments, and ways of making sense. But they *share* something: the boundary object. The specific meanings each of them attaches to this object are different. But as long as nobody stresses these differences, the boundary object doesn't seem to be two or three different objects. It remains fuzzy enough to absorb the possible tensions. It is a common object, shared by the various social groups. Thus, it facilitates collaboration across boundaries and thereby makes these boundaries less absolute. It blurs them.

Blurring boundaries is a way of contesting them. What is maintained in this, however, is the idea that there are different

cation and 96.8 among the controls. Look what this means: even if they had fewer heart attacks, the risk of dying was not smaller for those on medication but larger! Nobody knows how this higher mortality is related to lower cholesterol levels. But there it is ("Drie ton om een hartinfarct te vermijden," Volkskrant, 25 November 1995, p. 17).

A reduction in heart attacks of 31 percent is a good result. For the population. But the general practitioner interviewed isn't convinced that this outcome implies that his middle-aged male patients should now take cholesterol medication, come to checkups, and start bothering about their health. Their lives would not improve—ironically, their overall mortality might even increase.

What is good for individuals may be of little or no benefit for the very population they are a part of, whereas what is good for the population may have little or no value for its individual members. The interventions that improve the condition of individuals with atherosclerosis don't necessarily improve the condition of the population—and vice versa. This may be just a practical clash, framed in terms of economic priorities or as a question about how to spend scarce skills and energy. But sometimes the clash is shaped as a matter of principle. The general practitioner just quoted doesn't take prescribing cholesterol treatment to all middle-aged men in his practice to be "too expensive" or "too much work." It is simply bad for his patients.

And yet the improvement of individual and population also include each other. They clash *and* they include each other. Take the question whether or not PTA improves a patient's condition. When it comes to evaluating the PTA undergone by an individual patient (Mrs. Zoka again) she is personally asked how she's doing—or several of her body's parameters are measured with a test. But

---

regions. Adjacent to one another. With a lot of fuzziness between them, but separable, separate, all the same. This image has been contested in its turn. At this point I would like to relate to another body of literature (if only in passing) about spatialities in *geography*. Here, quite a while ago already, the model of the *region* for thinking about spatial formations was complemented with that of the *network*. (For an early and sharp articulation of this, see Lacoste 1976; for a later one, see Harvey 1990.) If people are followed in their daily movements, they do not live in a single

confined region, but in a variety of networks. The network in which they have telephone contact is far larger than that of the few shops where they buy their bread. The places where they go to when visiting family form a different network from those where they go for study or sports or holidays. And whether there are regional boundaries to cross or not is, most of the time, fairly irrelevant for moving through the network.

"Network" was also the term mobilized in the early eighties for understanding how science might be geographically situated.

when it comes to answering the question of whether PTA might be a treatment good enough to consider using it for individuals at all, population studies are designed.

The clinical epidemiologist has a room in the hospital. A colleague, three junior researchers. The secretary protects the entrance. She's scheduled an appointment for me. The situation of a researcher who comes to see the clinical epidemiologist is not unusual. People who are engaged in research often come to her for advice. And they get it. "Some doctors really know how to make a proper design by now. But still we're there to help them. And I must say, sometimes it's amazing. Sometimes designs are so bad. It will be some time before it's wiped out. Case thinking."

A treatment may or may not work out well in a specific "case." But where *the* treatment is good or not has become a different question. It can no longer be answered by pointing at (amazing or typical) individual cases. A treatment is only approved of if it has shown to improve a large enough percentage of its target population. Thus, population studies are included in individual treatment decisions.

But which population studies? Given the potential frictions between improving the health of populations and that of individuals, this is a difficult question. Trials, or so the clinical epidemiologist stresses, must be "properly designed." They must have controls and include a large enough population. Not the general population this time, but a well-delineated *target* population. What is also important here is which "indicators," "cutoff points," and "targets" will allow indi-

---

The practices of science are not confined to a single site, and yet the old idea that science is universal doesn't hold either, for it skips over the fact that we are dealing with *practices*. Practices are not everywhere: they are somewhere. Where? In Bruno Latour's version of this argument, we are asked to look to other networks in order to understand those of science. Take the camembert network, for instance. The camembert in the supermarkets of California is far away but no different from that in the supermarkets of Paris. They come from the same Normandy factory, due to networks of transportation and commerce that are able to cross state borders. Sci-entific experiments may, likewise, give the same results in Ghana as they do in London. But this only is the case if the laboratory in Ghana is equipped with the same instruments as that in London and staffed with equally well-trained people. As soon as there is a power cut in either place (and this happens more often in Ghana), the network is no longer capable of maintaining similarity. It fails. The question of whether Newton's laws are true in Ghana, then, does not depend on its distance from London in kilometers, but on whether steady electricity and some other crucial network nodes are persistently present (Latour 1988).

vidual improvement (that clinicians can see in their patients) to become visible in (the epidemiological assessment of) the population.

An internist: "This is interesting, you might find this interesting. We are facing the question right now about what to use as the indicator in our trials. We think that the risk of patients with diabetes of developing atherosclerosis is not only related to their sugar levels, but to their lipoprotein levels as well. Maybe even more so. So we are trying to monitor various lipoprotein levels. But what tells us whether or not this indeed reduces people's risks? Most studies of the prevention of atherosclerosis in patients with diabetes go by mortality from heart attacks. They do so for a long time now. But things have changed. Prevention has been reasonably successful. In a population of people who regularly see their doctor — and that means: people we can study — the risk has declined. It has declined, from say, 20 percent in a given time span to 5 percent in that same time span. But we think we may improve these patients' overall arterial condition more. If we are right, they'll have fewer CVAS [cerebrovascular accidents], less claudication, fewer renal failures. But it's unlikely that mortality statistics will be influenced all that much. Death from a heart attack — over a similar time span — might go from 5 percent to 4 percent. But it's almost impossible to get statistically significant data about such a small shift. You need hundreds and hundreds of patients. So we're looking for something else, something we can handle here in the hospital with the patient population we have. Some biochemical parameter in the blood, maybe. Some indicator that is sensitive enough to show on a population level what happens to our patients."

The effects of normalizing the lipoprotein levels in patients with diabetes may have become too subtle to be visible in mortality statistics. So counting death rates is no longer a good *mediator* between individual and population. Instead, some other way to enact atherosclerosis would be better included in the

Here, then, we have another genre for establishing difference and similarity. When the network holds, there is similarity. When it fails (when one of the alliances between the nodes gets disrupted or one of the nodes falls apart), then there is difference. So here the crucial transition point from similarity to difference is not a boundary, but the stability of the network elements and the concomitant functionality. A suggestion that, later on, starts to get blurred as well. Look at the diagnosis of anemia. In the Netherlands this draws on the measurement of hemoglobin levels. Clinical signs (like dizziness, tiredness, white eyelids) may be a reason to use the laboratory, but they do not replace it. In Africa, in some circumstances, the laboratory network doesn't hold. Looking at a tired, dizzy person's eyelids may be the only practice available to diagnose anemia. Does this turn the two "anemias" diagnosed into different ones? Or is the practice of clinical diagnosis, which can travel because it is incorporated in the body of a physician, a tenacious way of maintain-

population study. Instead of deaths avoided, another standard for improvement must be mobilized. A decrease in the number of people with angina pectoris or cerebrovascular accidents or claudication. Or some biochemical parameter. In order to become well established, the monitoring of individuals' lipoprotein levels must be backed up by a proper population study. But in order for this population study indeed to show the improvement the treatment brings to individuals, the appropriate parameters have to be found. They include each other, individual treatment and population study. And thus (for a practice to proceed fluently), each has to be delicately adjusted to the other.

Interference

In this book, so far, I have followed a single multiple object: *atherosclerosis*. There are many versions of it. A thickening of the vessel wall that happens to arteries *and* a cause of death that threatens populations. A stenosis to be scraped away *and* a process that, if attended to early on, might be slowed down or even prevented. And so on. But atheroscleroses do not simply relate to one another. In medical practice *disease* may be the central object of concern, but it is not the only one that is relevant. It is because disease is so central to health care, and understudied by outsiders, that I have unraveled a disease here. But for the praxiographer it is equally possible to follow other objects as they are being enacted. Take surgeons. Inside the operation theater, with all its sterile green, well-washed hands, and sophisticated instruments, a surgeon *is* someone allowed to cut in other people's flesh as if this were a technical, and not a violent, matter. Elsewhere this is different. A surgeon who takes out a knife in a decision-making meeting commits a serious transgression. Thus, *the surgeon* is no more one than atherosclerosis. Exempt from the taboo of violating other people's skin *here* (in

ing similarity from one place to another? Of maintaining the similarity of "anemia" that, despite failing laboratories, can still be diagnosed? To talk about such transient situations where there is *both* difference *and* similarity, where the precise moment where similarity turns into difference cannot be pointed out, where the transitions aren't clear, the term *fluid* was suggested. In the literature. A fluid space, then, isn't quite like a regional one. Difference inside a fluid space isn't necessarily marked by boundaries. It isn't always sharp. It moves.

And a fluid space isn't quite like a network, either. In a fluid, elements inform each other. But the way they do so continuously alters. The bonds within fluid spaces aren't stable. No single component—if it can be singled out—is absolutely necessary. But if they *all* fail, then that's the end of what existed. (Here another mode of relating to the literature may be deployed. A so-called see also my . . . But it is "our"! Whichever way, for a further exploration of this notion of the fluid, see Mol and Law 1994 and De Laet and Mol 2000.)

the operation theater), *elsewhere* (everywhere else in fact) this taboo holds for a surgeon as much as for the rest of us.

The example helps to elucidate why it is not a problem but an advantage that the term *enacting* leaves open *who* or *what* the actor is. Where originates the action when atherosclerosis is enacted? Many entities are involved: knives, questions, telephones, forms, files, pictures, trousers, technicians, and so on. But none of these are solid characters: after a little more investigation, all of them, like the surgeon, appear to be *multiple*. The surgeon enacting atherosclerosis as "pain on walking" may have the same face, voice, and name tag as the surgeon who "scrapes a thick intima out of an artery." But they differ. While the first is a talker, the second is a cutter. A single person may be capable of shifting from one repertoire (talking) to the other (cutting). The repertoires, in their turn, with all the materialities they involve, are capable of shifting the locally enacted identity of the surgeon.

Once we start to unravel ontology-in-practice there are no longer any stable variables. All variables vary from one site to another. The miracle to explain is how, even so, practices somehow hang together—the more so where variables vary in related ways. The enactment of a surgeon (as a talker or a cutter) and that of atherosclerosis (as pain or a stenosis) have something to do with one another. They are not complete dependents: a surgeon may talk about a "stenosis" or may, while cutting, express the hope the patient's pain will go away. But there are, even so, crucial *interferences* between one enactment and the other. It is during an operation that atherosclerosis is materially enacted as a stenotic obstruction in an artery, while a conversation in the consulting room is necessary to turn

Inclusions

In the subtext of this chapter I relate to a few texts out of the vast literature about similarity and difference. More particularly, I have related so far to the literature that tells how divides that mark difference may be contested. The divide between normal and pathological has been blurred with the notion of "problems"—problems that never have a complete solution but, in the best case, tend to suggest a few ways of dealing with them. Gaps between self and other are likewise blurred. Selves no longer sharply stand apart from, but are, in various ways, flowing over in their others.

Boundaries have been turned into fuzzy zones. And networks dissolve into fluids. However, one of the crucial arguments the chapter tries to make is not about blurring divides, but about the coexistence of what is markedly different. A specific form of coexistence is presented here: that of incorporation, of living as a part of what is other, or of holding what is other inside the self. Of *inclusion*. The literatures related to so far are relevant as a background to this because they situate the quest for articulating "inclusion" amid other ways of attending to practices of being different. But so far

pain from a problem nagging a person into a medically relevant fact. Interferences deserve a lot of further study. To give a first rough indication of where such study might take us, I give a small example.

It is the example of the interference between the enactment of atherosclerosis and that of *sex difference*. So far, in this text, I have attributed genders to the people who figured in my stories. They were *Mr.* or *Mrs.* So-and-so. They were *he* or *she*. But I have not gone into the relevance of these attributes to the reality of atherosclerosis — or vice versa. An omission that does not signal irrelevance, but rather a too overwhelming complexity of the topic. In the hospital, sex difference comes in even more variants than atherosclerosis, for it is almost everywhere. Jobs, working styles, professional roles, storage techniques, color schemes, bench heights, epidemiological tables, research questions, appointment hours, and so on: they all interfere with the reality of what it is to be *man* or *woman*. Thus, it is by necessity a gross complexity reduction if out of the resulting entanglement I pick one or two illustrative interferences between sex difference and atherosclerosis. I will try even so.

In the consulting room of hospital Z it is easy to find traces of all the clichés that have been made articulate in studies of sex difference elsewhere (plus some exceptions). Surgeons who want to find out to which extent the disease bothers people in their daily lives may ask a variety of questions. But invariably men of working age are asked about their jobs, and women of all ages about their household chores. The elderly surgeon is rough in a joking way with the young working class man. The young surgeon, when faced with a well-dressed, articu-

---

something is missing. Literature about inclusion.

It exists. But this image, that things *opposed* to one another may also *depend on* one another, is a rarely drawn one. This is because it directly counters common intuitions about what it is to be different, which are built on Aristotelian logic and a Euclidean conception of the space in which *positions* can be taken. The literature that breaks down these intuitions has indeed detoured to explore Aristotelian logic and Euclideanism. I'll point at two examples. The first is the work of Michel Serres (e.g., 1980, 1994). Staging himself as a philosopher knowledgeable about mathematics,

Serres ventures to depart from both Aristotle and Euclid. He makes stories that do so. His style is not argumentative, but seeks to feed the imagination. Arguments, after all, are Aristotelian in form. They rest on the opposition between A and *not-A*. But who wins most where two boxers fight, asks Serres? The answer is the person selling the tickets.

Don't attend to what is loudest, the fight, but shift your attention a little, widen it, and try to see what all this noise is part of. Opposing A and *not-A*, for instance, implies that A and *not-A* are relevant, meaningful expressions. It buys into an epistemic field in which they both make sense.

late elderly lady, relates to her in a polite and yet caring way that makes me think he is talking to a friend of his mother. Daughters who accompany their parents are questioned in more detail (and thus are supposed to know more) about their ailments than sons are. And so on. These differentiations are about what social scientists call *gender*. They enact social differences between men and women that do not directly depend on the sex of their bodies. However, with the possibility of a praxiography of disease comes that of a praxiography of the sexes. The study of enacted differences between *men's* and *women's* bodies. Medicine enacts bodies as having either one sex or the other. Or does it? This is to be investigated.

The pathology resident has finished her autopsy and moves into a small room with a large desk. She takes the little pieces of paper on which she has made notes out of her pockets and tries to fill in the autopsy form. It is a large form with a long list of preprinted anatomical structures about which she has to give her judgment. Skin: . . Liver: . . . . Lungs: . . . And so on. "Look," she says to me, addressing me as another woman who will be scandalized, but also embarrassed because today she is introducing me with some pride to her profession, "there are the genitals, here, preprinted. The penis, with all the details, glans and so on. And the womb is here too, but they have not preprinted the labia, nor the clitoris. So each time I've done an autopsy of a woman, I add them on, in pen. This form has been in use like this for ages."

It seems too crude to be true, thirty years after the first feminist criticism on medical textbooks for obliterating the specificities of women's bodies. What to do with such a finding, what can one say? It has all been said already.

Here is another snapshot story that at least allows for further analysis.

It blocks the exit to a world made up of entirely different entities. (An example: a discussion about whether to operate or do a PTA on arteries with a lesion longer than five and shorter than ten centimeters can be very fierce: *A* or *not-A*. But all along, operations and PTA are cast as good interventions and nobody asks about the possible indications for walking therapy.) What is opposed may also collaborate. But this, now, is not yet an image of mutual inclusion. To get there, what may help is the story of the bags.

Serres likes to point to simple objects, *things*, that are implied in our thinking. That are implied, not as objects referred to, but as models that inform the conceptual apparatus that we think with. One of the objects that he mentions is the box. More often than not, we take objects to be like solid boxes, relating to each other in a *transitive* way. A box is either bigger or smaller than another box. And if it is bigger, it may contain the smaller box—and if it is smaller it may be contained in the bigger box. But if, says Serres, we didn't cling to solidity so much, we might come to think about material, about cloth. About

For a single day I visit hospital Q. There is so much in Q that is different from what I've seen in Z that I give up my earlier idea of enlarging my field to two or three hospitals. One hospital is complex enough for my philosopher's aim of giving a frame to the study of ontology-in practice. More complexity would just make my story messy. But there is one small finding among my notes of that single day that I do not want to withhold from you. A meeting. Angiographies hang on a light box. With their fingers the radiologists point out the details: "Here, he has a bad stenosis, here, 90 percent I'd say." And so on. They keep on saying "he." One image after another. Only he. After the talking is over I walk to the files and check. Some of them are marked with an F, for female.

Here sex difference is neutralized away into masculinity. This may be taken to signal yet more disinterest in things female. But in this case more, or something else, is going on. In the decision meeting in hospital Z radiologists do differentiate between "he" and "she" pictures. But there this attribution doesn't make a difference to the assessment of angiographic images either. What is relevant about the lumen visible on an angiographic image is whether at the site of a stenosis it is, say, 70 percent or 90 percent smaller than lumen of the same artery higher up or further down. The stenosis of an artery has only to do with that artery: the sex of the patient to whom it belongs remains external to it. This not only has implications for arteries and stenoses (they have no sex) but also for the sexes: sexual difference is not filled in—fleshed out—with angiographic images of arteries.

In hospital Z (where I observed far more often) at some point in the discussion someone might bring in the patient's sex as a relevant fact. They might say: this artery is quite bad *for a woman.* This *for a woman* has no angiographic meaning. It refers to the epidemiological wisdom that the population of women is less

bags instead of boxes. If the blue bag is folded it fits into the yellow bag. But if it is taken out, then the yellow bag may be folded and fitted into the blue one. One may contain the other. And vice versa. They are, indeed, mutually inclusive. They relate in an *intransitive* way.

There is another example that also helps us to imagine mutual inclusion. This is the type of fractal in which two colors have been separated out—say red and green. But whenever one magnifies the red field it appears to have green specks inside it, and in the green field there are blotches of red. This is the image Dick Willems mobilizes (from a text of Serres) when he writes about the relation between patients and doctors. He gives the example of a patient with asthma and her general practitioner. Instead of taking the first to be all lay and the second to be all expert, he draws more intricate boundaries between them. Doctor and patient are different. But how? The doctor knows which doses of drugs to take against specific degrees of severity of the disease, and the patient doesn't for she

prone to atherosclerosis than the population of men. So *stenoses* may have no sex but their *incidence* does. This turns atherosclerosis into a sexed phenomenon (a disease that occurs more often in men than in women) and also specifically enacts the reality of the sexes (what *is* a man? Men are more prone to atherosclerosis than women). An enactment of the sexes that is challenged. As I told before, the professor of epidemiology of the medical school linked up with hospital Z was part of a working group bringing in *age* as another variable. The group argued that men have more atherosclerosis than women in populations below the age of sixty. As soon as entire life spans are taken into account, they claimed, the percentage of people who die from atherosclerosis is *higher* in women than in men.

Epidemiological sex differences may be contested. And contestation there may also occur when epidemiological sex differences, which say something about *populations,* are mobilized to say something about an entire class of individuals.

A right leg is made yellow with iodine. The genitals are blocked out of view with a green cloth. This morning's right leg is fat. That poses a problem for the surgeons. In order to find the artery they're after, they have to push the fat aside. They have to work their way through it along the entire track of the bypass they are inserting. At some point during this difficult work the senior resident says out loud (addressing the other male doctors present): "Hmm, I don't like operating on women. All this fat. It goes against my taste. And it makes me afraid of hitting on something unexpected, a nerve or something." The nurses, female, say nothing. But I catch one of them raising her eyebrows and looking exasperated at the others.

doesn't follow the latest medical literature about this. But the patient, in her turn, is an expert in using the peak-flow meter to find out about the current degree of severity of her disease, while the doctor doesn't have the necessary skills when it comes to manipulating this diagnostic device and breathing into it properly. Thus, the expert has expertise with lay patches inside it, while the lay person, however lay in some respects, elsewhere has appropriated patches of expertise (Willems 1992).

A second site in the literature where one can find such resources for thinking outside Aristotelian logics and Euclidean spatiality is the work of Marilyn Strathern.

Strathern has done fieldwork in Melanesia. In the present context the most intriguing aspect of her work is not what she *tells about* what she has learned there. It is when she actively mobilizes this in her own thinking. When she draws on it, draws it in. In doing so Strathern *practices* inclusion: her supposed object, Melanesian culture, has come to be situated *inside* her anthropological self. Its concepts have become part of Strathern's intellectual apparatus. Imagining and analyzing *inclusion,* then, is an important part of that apparatus. Strathern tells about differences that are not exclusive. She draws an image of how a son, contained in, may also contain the

The contestation is modest: raised eyebrows and a look of exasperation. But it is real enough. The surgeon moves from the epidemiological fact that *women as a population* have more subcutaneous fat in their legs than *men as a population* to a comment about *all women:* he doesn't like to operate them. Obviously, some women who make it to the operation table are thin, some men are fat. But the surgeon just quoted doesn't say that it is difficult to operate *fat people.* He doesn't say *fat* makes him nervous. *Women* do. He enacts sex difference as a difference in the amount of subcutaneous fat. The possibility to do so does not only depend on epidemiological tables, but also on the local reality of atherosclerosis. If atherosclerosis would not be enacted as a stenosis to bypass, a person's subcutaneous fat would be far less relevant to a doctor's efforts. And, thus, he would be less inclined to complain about it. And to blame the difficulty of today's operation on *women.*

Enacting atherosclerosis as a stenosis of an artery has *no* sex on an angiographic image, which implies that angiograms do not enter the piles of pictures marking the difference between *men* and *women.* Atherosclerosis does have a sex in epidemiological tables: at least for people under sixty atherosclerosis is more frequent in men than in women. At the same time the reality of what it is to be a *man* is marked by men's greater chance to get atherosclerosis before the age of sixty. In relation to the craftwork side of operations atherosclerosis has a sex again, another one this time: in the population of women there are more people with a lot of subcutaneous fat than in the population of men — which may be abbreviated too hastily as a dislike for operating on "women." Women are thus marked as having legs with lots of subcutaneous fat. These few examples

father. Of male and female forms that hold the other within. The other, she explains, isn't necessarily elsewhere. It may just as well be incorporated *within* the self. But it is incorporated without being assimilated. It is simultaneously here and other. Inside and different. To use her image: the other is a part even while it is not a piece cut of the same cloth (1992a).

Fractals also make their appearance in Strathern's texts. She comes with another fractal image: one that has to do with making parts out of a whole. Divide a thick line into two black stripes and a white stripe. So two-thirds of what you then have as your starting point is black, one third is white. Now divide each of the lines on your paper (or screen): the blacks into two black and one white line, the white into two white and one black line. And again. And again. The black and white lines get smaller and smaller. But however far you go, the amount of black and white remain the same. They only get more and more intimately included in each other. So here we have an image of a more and more ingrained *mutual inclusion* (Strathern 1991).

A move that, in some ways, academic texts relating to each other make as well. Texts that form *a literature* are all differ-

illustrate what it means to say that the reality of atherosclerosis and that of the sexes *interfere:* each of their versions may give a specific shape to some version of the other.

Ontologies

When objects are taken to be at the center of a variety of perspectives, the object world tends to be handled as if it were an assemblage of entities that hang together. That are part of one another. That cohere. But if we engage in praxiographic studies of the way reality is enacted, this transitive image of the relations between objects loses its appeal. Objects-in-practice have complex relations. An artery operated on is not necessarily smaller than the patient operated, nor is the first situated inside the latter. The artery may be bigger in that it receives more attention during the operation than the patient. And the patient does not contain the artery; he or she is not the body on the operating table *plus* something extra (a mind, or a social life). Instead, the patient is someone whom, at some other moment, the surgeon may exchange jokes with. It is someone who, elsewhere, may have a wife waiting for a telephone call. The two realities, that of the artery and that of the patient, do not encompass each other: they are, rather, situated *side by side.*

A corpse that lies naked and stiff on the metal table of the pathology department, waiting to be opened up, no longer breathes. It lacks life. But it doesn't need life added to it to be enacted as a person. A small cloth on its head, which may be taken away and put back again, is enough. And even after all the organs have been taken out of a corpse's abdominal cavity, the social life of the deceased may still be resumed. The caring hands of the pathology assistant are able to shift the corpse back into personhood again, by filling up its cavities, sewing the ribs together again, cleaning the skin that might be visible and, a crucial

ent, but also interdependent. They come to *include* each other. Thus, Strathern, just quoted, quotes, in her turn, the very text of Haraway that I earlier cited here: "My hope is that cyborgs relate difference by partial connection rather than antagonistic opposition, functional regulation, or mystic function" (Haraway, as quoted in Strathern 1991, 37). But while incorporating other texts, one alters them. While becoming included elsewhere, words acquire a different thrust, even while they stay the same. Strathern is articulate about this: "I have my own interest in Haraway's political cyborg" (38). Out of the layered meanings with which Haraway has infused her cyborgs, only some, altered, are mobilized in Strathern's work. Likewise, the relations I make to the literature are idiosyncratic incorporations of parts of other texts into this (novel) one. They make this text both parasitic on *and* contain as parasites what has been written elsewhere.

move, dressing the body. Clothes tend to be an essential part of sociability and of enacting personhood in public.

Beyond the praxiographic turn, the relation between objects is not hidden in the order of things, but enacted in complex practices. Thus, it need not be of a transitive kind. Artery and person are situated *next to* one another, rather than being a part contained inside a whole. A corpse does not become a person by *adding* life to it, but by carefully *taking away* and putting back again a piece of cloth. It may also happen that an object *is* enacted as a part of the other, but that the inclusive relation goes the other way around as well. To assess the severity of a disease in a population, individual instances are counted. But to assess individual disease, doctors take their epidemiological knowledge of its frequency into account. The population *includes* the individual—but the individual, in its turn, also *includes* the population. Sometimes this mutual inclusion may even lead to loops.

Mutual inclusion does not imply that there are no frictions left. The ontology of medical practice is not *the* ontology of a *single* practice: there are as many frictions between objects enacted as there are between the practices in which their enactment takes place. Aiming to improve the health of populations or rather that of individuals are goals that often are at odds with one another. And yet no population makes progress on any scale if no individuals' situations have been altered. And a treatment can only be established as good if it brings about a measurable change in a large enough number of people in its target population.

And then there is interference. This book unravels the enactment of a single disease in a single site: atherosclerosis in hospital Z. But this object is obviously not alone. It interferes with the reality of many others: surgeons, tables, pavements, X-ray, nurses, and so on. A few of the interferences between the reality of atherosclerosis and that of sex difference were mobilized here as an illustration. Coexistence side by side, mutual inclusion, inclusion in tension, interference: the relations between objects enacted are complex. Ontology-in-practice comes with objects that do not so much cohere as assemble.

It can be done. It is possible to write an ethnography of *disease*. This book shows
that this is the case. It has presented a patchwork image of atherosclerosis of the
leg arteries: a single disease that in practice appears to be more than one — without being fragmented into many. Thus, a body may be multiple without shifting
into pluralism. So instead of tracing paradigmatic gaps, this ethnography-of-a-
disease became a study into the coexistence of multiple entities that go by the
same name. In its turn coexistence comes in varieties and takes different shapes.
Here we have explored addition, translation, distribution (over different sites in
the hospital, different layers of the body, and different moments in time), and
inclusion. And if one begins to study the interferences between the enactments
of two or three multiple objects (such as atherosclerosis *and* sex difference), then
the complexities start to grow exponentially — though these are complexities to
be investigated elsewhere, for this is the point where this study stops. It has done
what it set out to do. A single/multiple disease has been described as a part of
the practices in which it is enacted.

But what is it to do this? What is done along with it? The stories in this book
do not finally unveil the truth about medical practice. Nor would I want to pose
as a member of a small avant-garde of theorists who finally know what ontology
is *really* about. None of this. Mind you, the stories assembled in this book are true
and in as far as they are not, they need to be put right. And I take the theoretical
apparatus mobilized and/or developed here to be worthwhile. But veracity is not
the point. Instead, it is interference. Like any other representation, this book is
part of a practice, or a set of practices. Attending to the multiplicity of the body

and its diseases can be done, or it can be left undone. It is an act. So in this final chapter I draw no final conclusions. Instead I briefly explore the act(s) this book engages in and point to some of those that it leaves undone.

How Sciences Relate

Shifting from understanding objects as the focus point of various perspectives to following them as they are enacted in a variety of practices implies a shift from asking how sciences represent to asking how they intervene. Over the past few decades many philosophers have stressed the importance of intervention as the dominant modern way of acquiring knowledge: epistemology lost its reverence for contemplation a long time ago. But, even if interference was important, interfering was not the point. The crucial issue in relating to objects was to get to know them. This book is part of a recent wave of studies that takes a further step away from disembodied contemplation. This means that it no longer follows a gaze that tries to see objects but instead follows objects while they are being enacted in practice. So, the emphasis shifts. Instead of the observer's eyes, the practitioner's hands become the focus point of theorizing.

Thus, this book contributes to a philosophical shift in which knowledge is no

---

Method

There is a large literature about method. Or rather there are three.

The first of these is of a legislative kind. It discusses how method should be shaped in such a way that the knowledge it helps to generate is *valid*. Valid knowledge should not contain the traces of the subjects who engage in knowing, nor of the situation in which the knowledge is articulated. It must be *pure*. No biases, no noise, should spoil a science's clear mirror image of the object. In this legislative tradition scientific knowledge should indeed be a mirror image of its object. The question of how this might be achieved is answered in a lot of different ways: very many legislative texts about method have been written. What holds this literature together is a quest after a method that is *good* in that it generates object-dependent, uncontaminated knowledge. (But what to refer

to? There is too much of it. No single title representative. But see, for example, Suppe 1977.)

The second genre in the literature is critical. It undermines the first. It tells that those who join the quest after a sound method have so far not found it. Along the way the main effect of their attempts at legislation has been to demarcate science from other kinds of knowledge. Such boundary setting has helped to protect some communities, those that succeeded in calling themselves "scientific," against outsiders. A large variety of examples are presented—not of method, but of the way it fails to keep out *bias* even though it is socially effective in keeping out *strangers*. Thus, we have come to learn about the manifest sexism contained in twentieth-century medical textbooks (e.g., Dreifus 1978). And about the subtle sexism, too (Jacobus, Fox Keller, and Shuttle-

longer treated primarily as referential, as a set of statements *about* reality, but as a practice that interferes with other practices. It therefore participates *in* reality. And various other shifts follow from this. One of these is that we need to reconsider the character of the relations between the sciences. Since the nineteenth century the various branches of science (physics, chemistry, biology, psychology, sociology) have been understood as differing not primarily in method (as was earlier the case), but in their objects of study. These were given by nature. They hung together in reality and ontology was the branch of philosophy that made this coherence explicit—often using the image of the pyramid. Each object domain was like a layer in a pyramid of objects ordered from the small and relatively simple to the largest and the most complex. And each science had the task of studying the entities in one such layer. Thus, at the bottom of the pyramid the smallest particles and the force fields between them formed the object domain of physics, and at the apex the complex social relations between groups of people were to be studied by sociology. One of the dreams that went with this ontological monism was that, in the end, full knowledge about the behavior of the smallest particles would explain everything else. Physics would explain chemical laws; chemistry would predict what happens to living bodies; biology would be able to explain psychological makeup and social relations. Not everyone agreed with this picture. During the twentieth century considerable effort has been devoted to establishing the existence of thresholds in the ontological

worth 1990). And many stories have been told about the way in which midwives and others were marginalized in the nineteenth century, when their skills and knowledges did not come to be taught in universities and thus were not granted the predicate "scientific" (e.g., Böhme 1980).

The third genre in the literature not only abandons the quest for a sound method, but also the critical campaign against it. Instead, "method" is turned into an object of inquiry. A variety of questions is being asked about it—in empirical mode. There are historical studies that go into the question of how the experimental method that is still with us got shaped and how it happened that so much faith was invested in it (Shapin and Schaffer 1985). Others wonder

why it was *method* of all things that came to stand out as the way of demarcating the scientific from the bogus (Dehue 1995). And yet other studies investigate scientific ways of working in an ethnographic mode: the sampling habits, labeling practices, ways of accounting, writing styles that may be found in present-day laboratories, offices, and scientific meetings. The knowledge that results from these ways of working does not mirror its objects. Do they fail to do so? But no. Mirroring is simply the wrong term. Passively rendering an object is not what science's systematic ways of working *do*. Instead, they actively constitute a traceable link between an object that is studied and the articulations that come to circulate about it. When

pyramid. Thresholds between dead matter and living organisms, which, unlike dead matter, can get ill and die. Thresholds, too, between biological facts about sex difference, skin color or disease, and social events that do *not* follow from these facts and therefore need to be spoken about in specific, social terms: gender, culture, illness.

In this order of things, knowing and talking about *disease* is both a task and the privilege of biomedicine. Chemists, even if they may know all about the molecules out of which cells are composed, cannot hope to explain the organism and its diseases. *Bio*chemistry is required, and it needs to include a pathophysiological branch. Medical practice meanwhile requires a further addition. For in order to attend to patients as *a whole,* biomedical knowledge of disease is not enough. The way people live with diseases should be attended to as well. In this way of thinking, "living with disease" was taken to be a psychosocial phenomenon called *illness.* Calls to attend to illness were often cast in critical language. Medicine was accused of prioritizing the physicalities of disease and neglecting its psychosocial aspects. But however harsh the criticism, it was built on a shared understanding of knowledge and the relations between the sciences, which was that knowledge is to be classified in terms of what it talks about and that these objects precede the knowledge. Body or mind. Disease or illness. Blood vessels or trouble with moving about. Biology or sociology.

However, if we come to the sense that knowledge is primarily about partaking *in* a reality, our understanding of the relations between the sciences also begins to shift. For whatever the relations between objects hidden inside the

moving from object to article we do not leave the material realm to enter that of theory and thought, but move, instead, from one sociomaterial practice (observation, experiment) to another (drawing, writing) (see Lynch and Woolgar 1990).

I separate out these three ways of relating to method here. They do not encompass all books that have been written on the topic—but leave some out that deal with different themes or ask different questions. And neither are the three ways separated here, separated out so neatly in libraries, at conferences, or in university departments. So there are fusions, gray zones, interferences. One of these is that

criticism of current methodological legitimations (style 2) feeds into the design of new methods—to turn these into *better* methods (style 1). This comes with hopes, for instance, that if the white male gaze is joined by female and colored optics, unbiased knowledge becomes possible, and objectivity is reached after all (see for a variant of this Harding 1986). In an analogous way empirical inquiries into the way science is practiced (style 3) are mobilized as a resource in writings criticizing methodological pretensions (style 2). If "method" is just a local, practical achievement, it cannot offer a guarantee that the knowledge that comes out of it is true. But

body—atherosclerotic plaque, peak flow velocity, increased cholesterol level—the practices in which these objects exist are concerned with a lot more with expensive or cheap apparatus, blood or flesh, forms or conversations, work hours, self-esteem, or insurance schemes. Treatment decisions are informed by the length of a stenosis *and* the length of a hospital stay. In practice, such diverse phenomena do not belong to different orders. It makes no sense to delegate them to separate layers of reality. They are all relevant and have to be somehow reckoned with together. What different sciences have to offer practice is different points of leverage, different techniques for intervention, and, indeed, different methods. One specialism may have dyes at its disposal, another knives, and a third the technique of humming, but in hospital practice they must somehow align and coordinate their objects.

However physical an intervention, the practicalities belonging to the so-called social are always and inevitably implicated in it. That is not to say that they are handled well. The quality of handling disease/illness and the rest of the world in hospital practice has not been the explicit concern of this study. But if a critic wanted to criticize physicians for attending poorly to, say, patients' experiences, the present analysis suggests a different way of framing this criticism. The point is not that in such cases some object remains outside medical attentiveness. It is rather that some intervention receives insufficient attention when

this reflects back on the empirical study of science: its own methods hold no guarantees either. Then what makes *science studies* better than the self-interpretation of scientists, or lay opinion? What are the grounds for its own claims to expertise (Ashmore 1989)?

An important question, but not one that has to be posed in this paralyzing way. What turning method into an object of empirical inquiry has taught us is indeed that no knowledge is beyond critique. Another method might have lead to different conclusions. Thus, there is no longer a formal reason to go with this, that, or the other product of science, however sound its method. But this comes with another shift, which is that knowledge should not be understood as a mirror image of objects that lie waiting to be referred to. Meth-ods are not a way of opening a window on the world, but a way of interfering with it. They act, they *mediate* between an object and its representations. One way or another. Inevitably. That means that it is not so surprising that the quest for a method for producing faithful representations took so long and that each time some critic was able to find biases that interfered with the objectivity of the results.

Studying methods empirically, then, generates another understanding of what they *are*. No formal guarantees, but specific mediators, interferences. The question to now ask is *how* they mediate and interfere. Donna Haraway has described an example that is illuminating in its exaggeration. It is a cage—a nuclear family apparatus—designed to study paternal love in monkeys. It was developed in the

medical activities are evaluated. In hospital Z, people with intermittent claudication are only considered for an operation if they report that their daily lives are seriously hampered. This implies that at this point operations are appreciated as a primarily *social* intervention. But this is not the case in studies that evaluate operations. Take the typical clinical trial comparing operations and walking therapy for atherosclerosis of the leg arteries. The list of parameters assessed will include "pain-free walking distance" but most probably not "actual weekly amount of walking," "changes in daily life," or "assessment of the intervention" in the patients' own terms.

How to attend well to the complex list of interventions that each medical activity entails? This question is left open here. But surely the first step is to consistently recognize that there are many entanglements in every action. To keep practicalities unbracketed. To treat everything in medicine as a practice. To engage in a praxiography. Praxiographic stories have composite objects. Disease is not different in kind to hospital stays or daily life. Each flows into the other. This means that the stories in this book are about disease itself just as much as they are about the practices in hospital Z that are intended to cure, alleviate, prevent, or investigate disease. The disease *as much as* the medical practices that intervene in it: the two go together. A microscope is used to look at plaque, while plaque, if it is to be practically relevant in a hospital, needs a microscope (and dissection, slicing, and staining techniques) to make it visible. Similarly, conversational skills (of both doctor and patient) and the complaint "pain when

sixties and seventies in the laboratory of Harry Harlow at the University of Wisconsin in Madison. Harlow first made "cloth mothers" and "bottle mothers" to test which of these offered the greater maternal love to monkey infants (who, faced with this awkward choice, preferred warm cloth over food bottles). Now it was the fathers' turn. "Each infant in the nuclear family apparatus, a planned social environment worthy of Disney Worlds, had access to the whole neighborhood, including his or her own father. 'Their parents, however, always remained home together'" (Haraway 1989, 240).

The nuclear family apparatus made it possible to isolate the variable "paternal love" as a specific behavior of male monkeys. This phenomenon wasn't available for study before the apparatus. The object wasn't lying there and waiting patiently. The apparatus delineated it. But if the monkeys hadn't responded so well, the use of the apparatus would soon have been abandoned. Did the monkeys respond well? "The fathers were nicely social with the babies and showed that they had a function in family life: threatening external enemies (experimenters mostly, Harlow recognized, in his always honest jokes)" (241). The nuclear family cage helped different observers ("experimenters") to make comparable reports. That was what it was made to do. But it did

walking" depend on one another. As do blood velocity and the duplex machine measuring it. And without the statistical calculations for extrapolating data from small samples there would be no at-risk populations on national scales.

This is why an ethnographic study may talk about disease. In the traditional ordering of disciplines, an ethnographer talking about disease transgresses the thresholds separating the layers of reality in the pyramid of objects. But the move made here is different. It is not a matter of turning the arrow round so that instead of the natural sciences explaining social phenomena a social explanation of molecules, cells, or bodies is being presented. Instead, another axis has been introduced, another approach taken: that of practice. The latter encompasses molecules and money, cells and worries, bodies, knives, and smiles, and talks about all of these in a single breath. Thus, it stands in an oblique relation to explanatory knowledge and the static pyramid of objects to which this refers. It approaches knowledge and object as parts of life, elements in a history, occurrences in strings of interrelated events. But no. To talk of an *oblique* relation is not quite right either, because this might seem to imply that the ontological pyramid, approached differently, is left standing as it is. But it is not. If practice becomes our entrance into the world, ontology is no longer a monist whole. Ontology-in-practice is multiple. Objects that are enacted cannot be aligned from small to big, from simple to complex. Their relations are the intricate ones that we find between practices. Instead of being piled up in a pyramid, they rather relate like the pages in a sketch book. Each new page may yield a different image, made with a different technique and in as far as a scale is recognizable, it may again, each time, be a different one. There is no fixed point of comparison.

more. It literally constructed the 1950s U.S. suburban nuclear family in a monkey version.

The point of stressing this is not to say observers should *not* interfere. They always do. In the same book Haraway beautifully shows how the ethologists who went to study primates out in Africa interfered as well. They pretended they were modest outside observers, who, by building no cages, left the reality of "their" primates untouched. But they made the animals *theirs* even so. They set up camps, appropriated the primates by giving them names in order to recognize them and communicate about them, arranged for them to get used to the observational presence of the ethologists, and so on. All this is not *bad* because it is interference. But it is interference. And the question of how to evaluate it shifts to a question of content. *How* does it interfere—and what to think of that?

Asking this question opens up a fourth and relatively new way of attending to method. A way that is normative again, and interested in the good: what is a good way of doing research, of going about the assembling and the handling of material?

The praxiographic approach allows and requires one to take objects and events of all kinds into consideration when trying to understand the world. No phenomenon can be ignored on the grounds that it belongs to another discipline. This doesn't make description easier. And since not everything can be held together in a page or two, other ways of delineating the world have to be found. Of course, there are many candidate traditions. In the present book I have built mainly on ethnographic techniques of observation and writing. But in various traditions of writing history, events have also been described with all their sociomaterial entanglements. It is no accident that history fails to fit into the ordered list of sciences where each is responsible for a slice of the ontological pyramid. History has always taken another entrance into reality. Another quite different but equally interesting resource for praxiography is found in the *material and methods* sections of scientific articles. In theory these specify as much as possible about the practices of investigation. They instantiate the recognition that the practices forcing an object to speak are crucial to what may be said about it. This recognition not only exists in written form but also resonates in interesting ways with the day-to-day self-reflection of medical professionals: a further resource for praxiography. In hospital Z, the death of a patient was always followed by a discussion in the staff meeting. The responsible doctor was required to describe the train of events that led to the patient's death. In this story, no particular "layer of reality" was privileged over any other. Deviant cells figured next to deviant dripping fluids; unexpected allergies next to the failure to check for them; heart problems were talked about in a single breath with names and

This time, however, the register in which *the good* is being played out has changed. Knowledge is no longer good in as far as it faithfully represents some object *as it is*. Objects do not slide silently, untouched, from reality into a text. Instead, there are cages or chairs, there is touching, asking questions, cutting up continuities, isolating elements out of wholes here, and mixing entities together a little further along. The new normative question therefore becomes which of these interferences are good ones. And when, where, in which context, and for whom they are good. Good knowledge, then, does not draw its worth from *living up to* reality. What we should seek, instead, are worthwhile ways of *living with* the real.

Self-reflexive desperation about the foundation of our (whose?) knowledge is no longer required. We would be wiser to spend our energy on trying to come to grips with what we are doing when crafting academic knowledge. What are we doing—when we go into fields, observe, make notes, count, recount, cut, paste, color, measure, slice, categorize, and so on. What are we doing when we tame materials, when we publish, give talks, stage stories for various audiences. Asking such questions means that we need to abandon the methods section of the library and

doses of drugs administered. If any actor was most central to these stories it was not the sick body but the speaking professional, for the leading question invariably was what, if anything, he and his colleagues might have done better.

The distinction between knowing *in* medicine (about disease) and knowing *about* medicine (that is about its practice) is blurred, not just in praxiographic studies like the present one, but also in historical studies, in material and methods sections, and in the hospital itself. Objects enacted and practices of diagnosing and intervening belong together. They are intertwined. They jointly differ from other object/practice constellations. The concomitant relevant axis of difference in the sciences, then, is no longer between the social and the natural sciences, or, more specifically, between *classes of objects* and the sciences referring to them. Instead, the axis of difference needing further exploration is between *versions of objects* and the (science-related) practices in which they are enacted. If a disease like atherosclerosis is more than one, it becomes relevant to ask which one "it" is made to be. Which one of its various versions is enacted at any specific site or in any particular situation? Is it an X-ray picture and the atherosclerosis that encroaches the arterial lumen; or is it a patient history and the atherosclerosis that gives pain-on-walking? Are surgeons operating on clogged vessels or are patients engaged in walking therapy encouraged by their physical therapist? This, then, is the crucial question in a world where ontology is accepted to be

move to the shelves that tell about the politics of academic work. Here I won't relate to that literature as a whole, but confine myself to what is on a single shelf. The shelf with the books that reflect on the effects of writing styles. (There is a lot on this shelf! But see, for example, Bazerman 1988; Trinh Minh-ha 1989; Clifford and Marcus 1986.) In different ways these three books tell us that what we are doing when writing academic texts depends not only on how the material is assembled. At least as important are the ways in which it is processed, rendered, mobilized. Written.

Is nature made to speak, or is a *materials and methods* section put somewhere prominently? Is "a culture" presented as if it existed out there, independent of the ethnographer who happened to come round to study it, or is it made clear throughout the writing that the stories told depend on scenes the author was a part of, even if it was only as an observer? Is the *subject* of writing staged as an observing outsider present in scenes she turns into "material," or rather as someone who approaches the field with fascinations, passions, and theoretical baggage that deserve a lot more attention than they get in methodological attempts to rule them out? (For general anthropological work, this is explored in, for example, Okeley and Callaway 1992. For a good example of what this may mean in science and technology studies, see Law 2000.) And what difference does it make if one presents one's study as a detective story, not just by using metaphors like "discovery" and "finding clues" but, more elaborately, by bringing

multiple: what is being done and what, in doing so, is reality in practice made to be?

Doubt

When I presented drafts of my articles or chapters of the many drafts of this book to my informants they were pleased when they recognized themselves and each other in my stories. Sometimes they suggested small corrections at points where I hadn't properly understood some technicality. Sometimes they nodded: yes, this is how it goes. But they also said they felt alienated. Somehow I made the familiar sound so *unfamiliar* to them. So strange. And yet one might say that hospital Z was not just the place where I assembled my material, but also a place where I learned a lot about the theoretical insights that I have presented here. For instance, the most concise way of articulating the idea that objects enacted depend on practicality was suggested by the resident who was my informant in the department of pathology. It was he, not me, who qualified his "this, here, is atherosclerosis" with the crucial *under the microscope.*

In hospital Z it happens time and time again that the practicalities of enacting a specific version of atherosclerosis are underscored. For instance, the techni-

this narrative plot to the fore and playing with it (as is done in Latour 1996)?

Texts are active. And they do so much. One cannot possibly engage in an explicit and articulate way with all of these activities in detail in any one text, all the more so if the text has something else as its core topic. Here, therefore, I've picked out a single stylistic characteristic to attend to. All academic texts somehow relate to the literature. The question I've posed to myself, and you, throughout this book, is how to do this. *How to relate to the literature?* By inserting a title. By presenting a quote. By relating a story. By giving one's text its place among others.

Rationality

When research is presented as requiring *method* in order to result in valid knowledge, the analogous recommendation for practice is that it must become more *ratio-*

nal. In a variety of ways this has been claimed and propagated over the past few decades: medical practice is too messy and in need of purification. The irrational should be washed out of it. There is a large literature about *how* to do this. Its quest for rationalization comes with the hope that scientific order can come to rule practice. There is a second literature arguing that rationalization shouldn't be strived after. Neat ordering isn't possible since practice has a specificity of its own, different from that of science. A third literature investigates *what exactly alters* when rationalization strategies are actively put into practice. It takes "science" as a set of practices as much as "practice" and wonders what happens in the interferences between different working styles.

The present study is intertwined with, or could be read as a part of, the third kind of literature. It helps to undermine

calities in *materials and methods* sections of articles also get a lot of attention in research meetings. "But in how many patients did you find that?" Or "Why did you measure pressures only in rest and not after exercise?" Or "What did you say you used as a calcium antagonist?" For the participants in the research meeting it is a truism that the outcomes of a research project follow from such details. They shape the facts. As long as there is reason for or an occasion to doubt, the technicalities of an investigation are kept in focus. As long as various roads may still be taken, the entire trajectory so far is kept into view. It is only once outcomes are accepted as facts that the methods by which they were reached are, at least for the time being, abbreviated, allowed to fade out, forgotten. The two movements seem to go together: the consolidation of a fact and the bracketing of the means of its production.

In the diagnostic process something similar occurs. If a doctor doubts the diagnosis of a colleague, then questions about technicalities are raised. "But *how* did you ask when this pain occurs?" Or "Your pressures are odd: are you sure the arteries weren't calcified?" Or "Who the hell decided to make an angiography of this patient?" Once an indication for treatment is written down, however, once there has been a conversation with the patient about it, and once the treatment is scheduled, such doubts tend to evaporate. It is on to the next task. A crucial bifurcation point is passed, the past is closed off, the practicalities of diagnosing are erased—all that remains of them is their results and a plan for treatment.

the presumptions of the other two, which both differentiate between scientific order on the one hand and mundane practice on the other. The praxiographic way to go about these issues is not to propagate rationalization strategies in general terms, neither is it to warn against them in equally general terms. Instead, it is to investigate what they bring along. What they do. It is to open up the question of *how* rationalization strategies alter what they interfere with. There are a lot of ways to handle this question. Here, I will present you with just a few examples of this third approach. They come from different places and each bring their own concerns with them but all tackle the question of the improvement of health care.

The first book on my little list is *Health and Efficiency: A Sociology of Health Economics* (Ashmore, Mulkay, and Pinch 1989). It pays a lot of attention to the question of how *health economics* manages to present itself as rational in the first place. How does it stage its capability of improving the way decisions in health care are being taken? How does it present itself as being able to help increase the (market) quality and decrease the (financial) costs of health care—all in one go? The authors state that the economists' claims of expertise are strengthened by their shifting between two versions: a strong one (that holds big promise and suggests that if its own economic rationality was obeyed things would get better) and a weak one

"Obstruc. fem. art. left; bypass to below the knee" or something like that. If, however, something unexpectedly goes drastically wrong at a later stage, it is almost always possible to go back in time. To take out the photos and have another look at them. To search the file for small traces that were missed earlier on. The treating physician who traces a history after someone has died is likely to look back into the past in this way. Was there a moment when we were sure about something, but we should have stayed a while longer in doubt?

Attending to practicalities also happens when some treatment is doubted. It helps to make space for other possibilities. An internist critical of operations on leg arteries said to me in an interview: "They [the surgeons] look at these angiographic pictures and come to think that they can *see* atherosclerosis. There it is: a pipe that is blocked and they need to unplug it." And then he added: "But by staring at an angiographic image one would never invent walking therapy." The image of the pipe that needs unplugging makes atherosclerosis into something unlikely to be improved by walking therapy. After all, walking does not unplug the pipe that looks so stenotic on the angiographic image. In an attempt to raise doubt about the necessity of surgery, the internist tries to undermine the reality effect of the angiographic images. Do not think that it is *atherosclerosis* you see there, but keep the specificities of the imaging technique in mind.

In this book I have argued that different practicalities of research, diagnosis, and/or treatment each address a slightly different "atherosclerosis." This idea is not alien to the hospital: I may even have learned it there. But there is a differ-

---

(that can be fallen back on in case of resistance: we know there are other matters to take into account as well). The authors analyze the contents of the economists' expertise as well. They look into the specificities of option appraisal, clinical budgeting, and the evaluation of interventions by assessing their (positive or negative) effects on people's quality of life.

Meanwhile the authors self-reflexively attend to their own claims of expertise. What is it to present one's stories as *knowledge about* health economy? In their desire to be serious about establishing a symmetry between economic expertise and their own sociological expertise, Ashmore, Mulkay, and Pinch have written a book that

is full of mockery. (At this point I must insert a remark. An aside. However much "writing" has become a topic that is theoretically discussed, there still aren't many books that *do* something to enrich, complexify, and change academic writing practices. Writing methods are still not taken as seriously as methods of gathering and analyzing material. *Health and Efficiency* is among the few exceptions. It brims with conversations, shifts in scenery, alternative presentations of material, self-reflexive remarks, and jokes. How to relate to that? In awe or with mere admiration?) So. So claims of expertise are robbed of their foundation.

The issue is not that health economics

ence. In hospital Z the other repertoire exists as well: that of bracketing practicalities. That of speaking about atherosclerosis *tout court* without mentioning microscopes, interview techniques, angiographies, or any other modality of enacting the disease. Of atherosclerosis in isolation. At such moments what one might think of as a virtual common object is projected into the body, an object that is hidden underneath the skin. An object that may be approached in various ways, that shows a variety of aspects, but that in the end is one. There it is, and suddenly it no longer seems to be a part of practice, but a referent in a pre-existing reality: overwhelmingly real. At such moments doubt is smothered and certainty is being manufactured. "But surely we are all fighting the same disease? We share a goal, don't we? Obviously we all want to improve the health and lives of our patients." At such moments someone might say (to me, for instance, in reaction to this text): "But listen, people die, people suffer. There is a *real* disease out there." As if the certainty of death and misery necessarily brought with it the singularity of the real.

So in the hospital there are, at least, these two repertoires. Keeping the practicalities of enacting disease visible so that what happens may be doubted, *and* bracketing practicalities while working along and being confident in doing so. Making space for other enactments of reality, other versions of the disease to

should seek a better foundation from now on. "No, it is not the epistemological status of applied economics in any abstract sense that concerns us but rather the specific moral and political implications of its underlying assumptions" (187). If the authors have problems with health economics, its lack of rationality is not among these. The point is that in various instances interferences are made that could have been made otherwise. Had this been so, other outcomes would have followed. These are problems of *content*. An example. The QALY is a *quality-adjusted life year*. It is added to earlier epidemiological assessments that measured only survival. The addition is that the *quality* of the years patients survive an intervention are taken into account. But how? The QALYs obviously do so in a specific manner. One that allows accounting. One that

fits into the forms of quantitative studies. One that supposes that "aggregate data on preferences correctly represent the individual evaluations from which they originate" (192). Ashmore, Mulkay, and Pinch point out that sociological investigations into people's appreciations of their life could also proceed differently.

However, Ashmore, Mulkay, and Pinch do not develop an alternative health economics. They cast doubt since their primary concern is with the arrogance with which economic expertise is presented as lying beyond doubt. A predicate of scientificity is used to close off discussion. The economists put themselves *above* the practice they aim to improve. An extensive quote here, for Ashmore, Mulkay, and Pinch have put it in a way that asks to be quoted. (In relating to the literature one comes across this style characteristic:

be diagnosed and treated, or closing off alternatives so as to move ahead on a given track. Doubt and confidence: in the hospital they alternate. My informants know how to shift between them; I abstain from doing so. This suggests that the strangeness of this book lies not in its novelty, but in the persistence with which it never comes to rest in a sure, single, mortal body, but keeps on pointing at the practicalities of living. Being so stubborn is like remaining in doubt. An analysis like this opens up and keeps opened up the possibility that things might be done differently. Look, they *are* being done differently: a little further along. If something is self-evident here, then in that other site and situation it is not. If atherosclerosis is a thick vessel wall here (under the microscope), it is pain when walking there (in the consulting room), and an important cause of death in the Dutch population yet a little further along (in the computers of the department of epidemiology). Reality is varied.

In stressing ontological multiplicity this book lays bare the permanent possibility of alternative configurations. The doubt that might lead there isn't always practiced, but it *may be*. Medical practice is never so certain that it might not be different; reality is never so solid that it is singular. There are always alternatives. There is no body-isolated that may offer us a place beyond doubt. But this means that no version of atherosclerosis should necessarily be practiced "because the body itself leaves us with no alternative." Bodies enacted are being

some texts are *quotable*, while others, even if they are well written, are not.) "Efforts at reform and change must, and will, continue. Applied social scientists of all kinds will continue to make a major contribution to these efforts. And as they do so, they will, like the health economists, be faced with the fundamental problem that the very practices they wish to alter will tend to frustrate their efforts. The point we wish to emphasize is that confronting this 'problem,' if it is understood in the way we suggest in this book, is the essential first step towards a better form of practice (if we may be permitted such a blatant evaluation): one that consists of a willingness to work with, rather than against, the actors in the domain of application; one that is collaborative rather than imperious; modest rather than megaloma-

niac; and wishing to learn rather than itching to instruct" (195).

This literature link is strong. The present book presents a very different study, but it leads to the same conclusion (or has this been one of its driving forces, a conviction that informed this study all along?). If there are so many *rationalities* in practice, in the plural, mixing with one another, interfering, then why present oneself as an outsider, who, with a single mode of ordering, may change everything? Why do so as a rationalist, a radical, a revolutionary, a rightist of whatever kind? The tenacious character of such hopes is all the more surprising when one looks at what happens, in practice, with rationalist schemes when they are introduced to a specific site or situation. It never happens that everything gets subsumed under the newest head-

done, which means that they cannot answer the question *what to do.* However uncomfortable this may be, this question, what to do, is a question *we* have to face. Not in circumstances where anything is possible, but still. Reality used to be a standard to live up to, but given the proliferation of technoscience the question that now needs asking is "what reality should we live with?" That means that reality moves. It can no longer play the role philosophy cast for it a few centuries ago, the role of something to get in touch with. The role of something to grasp. To hold on to. To be sure about. The crucial philosophical question pertaining to reality was: *how can we be sure?* Now, after the turn to practice, we confront another question: *how to live with doubt?* It isn't easy. But somehow we must come to terms with the fact that we live in an underdetermined world, where doubt can always be raised. Somehow we must learn to understand how it is that, given this possibility, we can still act.

This, then, how to live with doubt, how to live in an underdetermined world, is another question that this book leaves open. However, part of the answer lies in shifting repertoires when considering action. If the question *what to do* no longer depends on *what is real,* then what else might it be linked up with? I suggest that if we can no longer find assurance by asking "is this knowledge true to its object?" it becomes all the more worthwhile to ask "is this practice good for the subjects (human or otherwise) involved in it?" If faithful representations no longer hold the power to ground us, we may still seek positive interventions. Thus, instead of truth, goodness comes to the center of the stage. Or rather,

ing. Instead, one more mode of ordering is added to the many others that are already there. This is what we learn from the next book on my list: *Rationalizing Medicine: Decision-Support Techniques and Medical Practices* (Berg 1997). This book tells a few stories. A first is that rationalization strategies may claim to *improve* medical practices, but the standards by which good and bad, and thus "improvement," are assessed do not precede them. They are, instead, framed in the process of developing and introducing the rationalization strategies—with which they are inextricably linked up. A second story told is that the opposition between a messy practice and a single rationality that comes to

its rescue does not hold because there are serious incongruencies between the various rationalities involved. Computer-based diagnostic tools incorporate a quite different rationality than clinical decision analysis. Protocols are different yet again, and so are expert systems.

And then there is a third story in this book, one that says that when it is introduced into a practice an ordering device does not *expunge* messiness, but shifts it. Pushes it along. An expert system, for instance, may solve some problems, but creates others. It may suggest useful interferences between the data it is fed with and a diagnosis, but it obliges the people working with it to feed it with data and

not *goodness,* as if there were only one version of it, but *goodnesses.* Once we accept that ontology is multiple and reality leaves us in doubt, it becomes all the more urgent to attend to modes and modalities of seeking, neglecting, celebrating, fighting, and otherwise living *the good* in this, that, or the other of its many guises.

A Politics of Who

The recognition that medicine is entwined with *the good* has led to the call for what is sometimes called "patient autonomy." Rather than professionals, "people themselves," "patients," are to decide what is good for them. Their norms are to be given weight. They have to make the judgments, and the role of professionals is simply to present patients with the options. Patients choose. A large industry (of literature, conferences, and committees) has grown to specify how to implement this requirement. What if the medical ideal of benefiting people clashes with the ideal of granting them autonomy? Are there moments when a professional should step in and decide for a patient? What to do with patients who are not capable of expressing their will in an articulate manner? Ignoring the complexities of such issues here, I want to stress that the growing interest in medicine's normativity has predominantly focused on *who* questions. Questions about *who* is being put, or should be put, in the position to decide what counts as good.

There are, roughly speaking, two ways in which "patients" are put in the position of making supposedly crucial normative decisions, two ways of living out a right to choose. The first is that of the market. Here, medical interventions are

then adapt these where they do not fit. What, for instance, if the system wants one to locate pain in the back or the front, while a patient tells about pain that moves from one place to the other? Practitioners working with systems that want to be fed with discrete information must constantly negotiate their fluid findings. And it is, finally, story four that tells that while decision-support tools claim to simplify practice, in fact they do not do so. They introduce, and thus *add on*, a further logic to those that are already there. Something that is likely to complexify practice yet further. This is not an argument against decision-support tools. A hammer may also complexify building practices and yet be a welcome extra tool. The question to pose, however, is what it might imply for designing tools. How, or so Berg asks, to build tools that help to improve practice, without fantasizing complexity away? Again a question that resonates with the stories I've been telling here. Improvement and rationalization are not quite the same.

The third book on my list is a sociohistorical case study. The case studied is that of the *clinical trials* set up to asses the value of drugs against the human immunodeficiency virus in the United States.

displayed as if laid out on a counter. Professionals turn into sellers who supply a product plus the information that allows the patient-customers to choose between the products on offer. The patient is to make the value judgments—even if to some extent everything on offer in a market is, by definition, a *good*. In a market, goods that are entirely worthless are supposed to disappear; there is no demand for them. Even so, the health care market is heavily controlled. Professionals are required to be licensed and to check the quality of their own and each others' products. Although in actual markets money is central, the crucial element of the market metaphor in the context of medicine is that the individual patient, being the customer of health care, is the actor of an individualized choice for or against some "care act" or isolated "intervention." An ideal patient-customer is able to find the goods that fit his or her specific needs and situation.

The second genre of handling choice is civic in character. Here, medical interventions do not figure as goods on a market, but as policy measures. They are interventions into modes of living—configuring professionals as kings rather than hawkers. The civic metaphor tends to turn the patient into a citizen who deserves to have jurisdiction over the interventions into his or her own body and life. Decisions have to be singled out, and the patient must then be able to argue civically for one course or action or another. But the civic metaphor doesn't necessarily argue for individual choice. Intervening in one life, after all,

The book is *Impure Science: AIDS, Activism, and the Politics of Knowledge* (Epstein 1996). It underlines the fact that clinical trials depend on the cooperation of many —the patients not the least among these. In the trials for drugs against HIV, some of the exigencies built into local definitions of good universal science were incongruent with the way most patients perceived their own interests. In the United States, getting enrolled in a trial may be of direct interest to patients because it provides them with free treatment. Moreover, being enrolled in a trial was often the *only* chance of receiving an antiviral treatment at all. But patients were obliged to refrain from taking drugs other than the one being tested in the trial. For some-

one with AIDS who has opportunistic infections, this is an unreasonable demand. Epstein describes how ACT UP, a movement of patient advocates, came to voice this and similar issues. First, they made their voice heard in angry protests against the way trials were designed. And then, as a next step, they were invited to come and sit on the committees that designed the trials.

The designs were adapted. At various points there appeared to be elements to contest. The question of who was included: at the outset participation was limited to (mainly white) males who had been infected through gay sex. It was only after ACT UP protests against this that first women were included and then drug ad-

also influences others. This brings with it the requirement that individual decisions should not harm others. But where does harm start? If one person chooses to have offspring through in vitro fertilization, this alters the meaning of being childless for others; if one would-be parent chooses to abort a fetus with Down syndrome, this touches on what parenting a child with Down syndrome does to others. To account for the effects of policy measures into a single patient's situation on the situations of others, the civic metaphor has been further developed. Interventions are understood as a way of organizing not just individual life, but that of the entire *polis,* the *body politic.* They concern us all, as patient-citizens. This implies that in the civic version of the *politics of who,* "patients" must represent themselves whenever decisions (about the organization of health care, allocation of money, research efforts, and so on) are being made.

The market genre and the civic genre have a common concern with the question of *who* decides. Both genres are informed by the same suspicion of professionals who patronizingly decide what is good for the rest of us. Ethicists together with social scientists investigating health care have contributed greatly to the articulation of this suspicion and have stressed the importance of the question of who decides in medicine. They have contributed to a politics of who. But

dicts (more often of color). The rules about taking other drugs were altered; with some adaptation of the statistics used this could be done. Then there was another intriguing issue: what should be taken as a parameter to mark the success of treatment? The primary choice of the epidemiologists had been to count the number of deaths in the treated and in a control population. But when survival became a little more prolonged it was argued that this was too slow. An intermediate parameter, a T-cell count, was chosen, a measure that was later discarded again. An appropriate parameter was difficult to find. What is interesting in the light of the present book is that in this specific case it became clear to all those concerned that what makes a parameter *appropriate* is a complex question. Statistical issues, the immune system's behavior, patients' hopes and expectations, health care finances, the pharmaceutical indus-

try's research style, government regulations—all of these are intertwined. The loudness with which the various elements are heard may differ depending on the specificities of the situation. And in this case, a patient-advocate movement willing and able to engage with the details of the science involved was crucial.

There is still a lot to learn about such engagements. What they require, for instance, is that the professionals involved allow themselves to be addressed, that they listen to what others have to say and take it as an argument to be included in the accounts. What they also require is that the "others" in question are able to mobilize the arguments with which to engage in such "addressing." Epstein stresses that the ACT UP people were highly educated —if in different fields. They didn't take long, moreover, to educate themselves in clinical epidemiology as well. This is not

such a politics of who has some problems. The first is that, although the customer and the citizen may be protected against such things as monopolies of suppliers or the power of the state, their will and their desires are supposed to be set, predetermined, and clear. The analyst takes the position of a lawyer for the patient movement whose task it is to make space for the patients' silenced voices. But the position of the lawyer is not the only possibility. What if the analyst takes the position of the patient himself or herself? Then it may well be that other matters become important. For instance, "how might we gain the right to decide" may be displaced by the at least equally urgent question "what should be done?" What might it be good to do? What might the good *be*, here and now, in this case or that other? The problem, then, is that in trying to give "the patient" a say, a politics-of-who remains silent about what, if one is a patient, one might actually say at the crucial moment.

The second problem with a politics-of-who is that it isolates moments when a choice is being made. It separates decision-making moments from the series of long layered and intertwined histories that produce them, as if somehow normative issues could be isolated and contained within those pivotal points. As if they were, indeed, pivotal points. Take the situation of a consulting room where a decision is being made about whether or not an operation would be a good thing for the patient who has come to seek help. A decent doctor would explain quietly about what is wrong with the patient's arteries and about the pros and

a story in which assembled experts were confronted by individual lay people who were only knowledgeable about their own case. ACT UP activists draw their insights from many people involved. They brought their own expertise along. Expertise about the daily lives of patients, to begin with. This allowed them to help build interferences between the daily lives of patients and the exigencies of doing clinical epidemiology research.

Thus, however much Epstein's story starts out from a sociological curiosity about the way lay people came to speak *inside* science, the lines of difference tend to be more complex than lay/professional divisions. For epidemiologists who had been involved in cancer research, for in-

stance, the ACT UP points were less alien than for those who so far dealt with acute infectious diseases. The committees designing trials welcomed the expertise on patient's concerns and daily lives ACT UP brought along. Without this they knew they risked setting up studies nobody would want to participate in. And ACT UP people in the end became so involved in the clinical epidemiology that they in their turn found themselves confronted with outsiders in the movement speaking out in the name of daily life. So the *who* question weighs heavily in Epstein's sociohistorical account. He persistently asks questions about who speaks and who does not. But what Epstein also makes clear is that once they were listened to, all those in-

cons of an operation. But to concentrate on *this* situation hides many others. For instance, that—at least in the Netherlands—the patient is present without having to think about the costs of various diagnoses and treatments. Or that no structured walking therapy was ever offered and that, despite huge investments, no drug is so far good enough (maybe the grant application that would have led to the drug was turned down—so what about *that* decision?). Or that some other hidden factor made this patient's atherosclerotic process go so far as to give pain on walking, so why wasn't that process ever intervened in? Or how come the patient had not considered this pain (as others might have done) as simply a part of getting old? Every single moment always hides endless contingencies— which, if we look at them carefully, are likely not simply to be contingent. That means that most elements relevant to making or unmaking the *goods of life* involved in making a decision escape the moment of that decision.

The third problem with a politics-of-who is that it is designed to push the power of professionals back, claiming some choices, and then more and yet more, for patient-customers or patient-citizens. But this same politics of who has trouble getting inside professionalism. It does, after all, grant professionals the facts. It requires of them that they give information—as if, from the beginning, there were a neutral set of data to lay out on the table. But there is not. My informants in hospital Z would stress that however much they tried to give "neutral information" they always found that the way they presented the facts made an impressive difference in how these facts were evaluated. But there is

volved, professionals and lay, preoccupied themselves primarily with another question: *what*. What is important, what should be done? Actors who have gained rights to speak no longer worry about getting heard, but wonder what to say. Maybe it is a matter of time: one question is not in tension with but follows after the other. If so, I would like to mobilize Epstein's book here as a support for a claim. This claim. It is time, in health care, to assemble and develop the theoretical repertoires needed for a politics of *what*.

Locality
Where do texts come from and where do they go? What place or places do they carry along or within them? If we think of the present book, this question comes in various forms. One is that the material mobilized here may be situated as stemming from what in *anthropology*, despite energetic debate, is still called a culture. The way professionals and patients described here behave, calmly carrying out conversations, for instance, could be called *Dutch*. And so could the primarily clinical orientation of "my" vascular surgeons, highly consequential for what I say about the character of medicine. A second, *sociological* typification of the provenance of the material I have explored would be quite different: many sociologists would say that the object that I describe is micro. It is local in

more. Which facts should be presented? Which facts are pertinent to the reality of atherosclerosis: those of pathology or those of the clinic; hematological or epidemiological facts; duplex graphs or angiographic pictures? This is not simply a matter of which textbook page to turn into a nicely illustrated, suitably didactic leaflet. It is also a practical issue. Which machine to put to use, with what money to pay for it? Which hurts to evoke and which casualties to risk? Information, presenting some version of reality, does not come after practice. Neither does it precede it. Instead they are intertwined.

This book does not try to show that *the social* is larger than we took it to be while *the technical* is smaller. Instead, it suggests that technicalities themselves, in their most intimate details, are technically underdetermined. They depend on social matters: practicalities, contingencies, power plays, traditions. Thus, technicalities should not be left to professionals alone. They affect us all, for they involve *our* ways of living. But this does not mean that they are not also technicalities. That is why the present book does not try to push back the role or the power of the medical professionals a little further by revealing further patches of the medical domain where values exist alongside facts—and where, therefore, laypeople should make the decisions. What if values reside inside the facts? Then it may be better to stop shifting the boundary between the domains of professionals and patients and instead look for new ways of governing the territory together. But this suggests that ethnographers, philosophers, and soci-

the sense that it comes from somewhere small. The big picture is not sketched out. The social organization of health care, long-term developments in the biomedical sciences, the distribution of power, the flow of capital, what have you—all such macrophenomena escape the microsocial framework of the book. And then there is a third possibility. The *philosophical* tradition situates texts differently yet again. It does not link them to their places of origin, but rather to their destination. True philosophy, or so the dominant tradition suggests, comes up with *universally* valid theories. These transcend the specificities of any single site and move everywhere, without transportation costs. And since this book has generated no universalities, it would be said to fail as philosophy. If it deserves

to be taken seriously at all, it is as *mere* social science.

So we have three different modes of localizing. Let's look at them in a little more detail. First, then, culture. There is the question of the so-called cultural specificity of the events that take place in hospital Z. Are they distinctively and locally Dutch? One of the reviewers of this book, a North American, kept on insisting on this. With an ocean between us, he or she saw *Dutchness* running as a thread through every page—and challenged me to acknowledge this. So what to say about this? First, yes, there is a topic here. But second, it is one that deserves its own investigation. What might *Dutchness* be? In the local bookstore I found a book on the topic, a book that draws together a

ologists of medicine as, or just like, patients need to explore and engage with professionalism. Once inside the hospital, the *who* question is linked to, or even overshadowed by, *what* questions. There, time and again, the question to share is: *what to do*. What to do, this is a question *we* face, and the "we," or so I would want to argue, should be taken as widely as can be. But what kind of resources do "we" need if we are to face this question? Framing languages and shaping practices for dealing with the question "what to do" is part of a *politics of what*.

A Politics of What

For the medical profession, *what to do* has always been an important question, indeed recognized as having a normative dimension. However, the norms involved were naturalized. Saving lives, improving health—that was what medicine set out to do. The value of life and health was deemed to be given with our physical existence and beyond dispute. When patients may die of pneumonia if it is not treated and survive when an antibiotic is prescribed, no further questioning of the normativity of such treatment seemed necessary. And if insulin postpones the imminent death of patients with diabetes for decades its *goodness,* again, is accepted as obvious. When it is clear that the overall health of the population improves when people do not smoke, then warnings are printed on

lot of anthropological field studies done in the Netherlands. Not coincidentally, it is a book in Dutch (van Ginkel 1997).

The book situates the beginning of the anthropology of the Netherlands (beware: *Holland* is only a province of this country!) in an American text. It is a text written by Ruth Benedict in 1944 while at the Office of War Information in Washington. Dutch anthropologists had been active for decades in Dutch colonies, studying villages in Java, irrigation in Bali, rituals in New Guinea, and so on. The aim was to bring such places closer to administrators, merchants, and planters. But there was no need to bring home *itself* closer—for the Netherlands were nobody else's colony. But then, in the Second World War a lot of U.S. soldiers were to be stationed in the Netherlands. In order to reduce friction between the soldiers and the Dutch popu-

lation, each group had to learn about the other. And this is why Benedict assembled whatever written material about the country she could lay her hands on and sent out students to interview Dutch immigrants. With this material she wrote an exposé of *the Dutch character,* not so much stressing our obsession with clean houses (something travelers had remarked on over the centuries) as the self-assuredness of the Dutch. The Calvinist majority in particular, Benedict wrote, is convinced that it has Right on its side. Quote: "One can fairly say that the typical Hollander is so sure of himself that he does not submit to dictation. He stands up for his rights. He hates sentences beginning 'You must . . .' A so-called true story illustrates the Dutch attitude: The postmaster asks a little boy at the stamp window, 'What must you?' (a colloquial phrase). The little boy answers,

packages and doctors tell us not to smoke. Medicine never concealed its normative character. But its self-reflection was not directed at its central goals: postponing death and improving health. It became the profession's central concern, instead, to see if its interventions indeed helped to achieve these goals. Since roughly the 1950s more and more *clinical trials* have been conducted to evaluate which medical interventions succeed and which others fail to bring about improvement. Clinical trails have become the dominant mode by which the value of interventions is judged by the profession.

Trials, however elegant instruments though they may be, are not sufficient if we are to engage in a *politics of what*. They were designed in a time when, indeed, the goals of medical intervention were taken to be given with the natural characteristics of the body. Survival and health. At some point these goals have proved to be insufficiently specific. The first difficulties arose in cancer research. As long as "survival" is accepted as a goal, a treatment for cancer may seem successful if those who receive it live, say, an average of six months longer than those who do not. However, the patients and the physicians and nurses engaged in their day-to-day care weren't always convinced that such "survival" entails an improvement. Six months in and out of a hospital, with a disintegrating body, with pain from both disease and treatment, may well lead to more suffering than relief. In the discussion that followed, the goal of "survival" lost its self-evidence. Maybe it wasn't a natural good after all. Maybe the extra life treatment may bring was only a good if it was spent well, if the months gained

'I must nothing. But you must give me a stamp of two cents'" (Benedict 1997, 226).

So maybe it is no wonder that I show resistance when a reviewer presses me to attend to the Dutchness of my text. Must I write on Dutchness? Oh no! I must not! (My Dutch character has arguments to support it, too. After all, only *exotics* are required to culturally localize themselves. And, one may wonder, what kind of imperialist power [benevolent or not] hides behind the interest this time?) But more is going on here. *How* to account for, how to typify *Dutchness*? Attributing a character to "a typical Dutch person" may have been useful to the writers of a pamphlet to be dropped from airplanes to inform the in-

habitants of the country about the foreign soldiers. It may even have led to an instructive leaflet for the soldiers who needed to realize they differed from the natives. (The crucial point being that they were not to expect the prudent Dutch girls to be inclined to have sex with them.) But in most other contexts it has little pertinence. A large half century later anthropologists no longer tend to delineate national cultural characters at all. Anthropology has shifted from this to the investigation of patterns of shared meaning, and then on again to other ways of articulating similarities and difference.

Notities over Nederlanders lists a variety of ethnographic studies done in the

were indeed worth living. The term *quality of life* was coined to fill the gap left by many people's disappointment with survival alone.

So in current practice, clinical trials assess medical action not just against physical parameters, but also compare the impact of treatment on people's quality of life. Another step, maybe, toward a politics of what, but there are many more to take. For instance: in the quantitative research tradition of the trial, the question about what gives life quality and what does not is still taken up in a quasi-naturalizing way. A sociologizing way, or so one might say. What *the good life* might entail is not recognized as an essentially contested and thus a political issue. Instead, research is set out in a way that objectifies this good. Surveys are used to record individual opinions, weights are attributed to these, and they are entered into statistically sophisticated accounts. In this way quality becomes a quantity. Values are turned into facts, social facts. All the controversies around the question of what a *good life* might be are stifled. That people have different investments in life, that we clash when it comes to striving after the good, is turned into a mere calculative challenge. We are each accorded our own opinion. Here, fill in your form, of course your opinion will be taken into account. Not as a political act, however, but as a social given. Instead of being staged in a theater of discords, differences are flattened out onto a spreadsheet.

If I advocate a politics-of-what here, it is not to suggest that the state should get involved in every detail of what happens in the hospital by proliferating laws. It is, instead, to stress that all these details involve "the good life." Relating this to clinical trials, one might say that not only issues now categorized as relevant

Netherlands. Some unravel the fishermen's trade, others stem from field work in orthodox Protestant villages, yet others follow heroin users in Utrecht or boy prostitutes in Amsterdam. They all explore the specificities of these different sites and situations. However, if we take them together they make it more difficult rather than easier to answer questions about *Dutchness*. What might these sites and situations have in common with each other? What do they share with hospital Z? Some works are presented that report on care practices, the most intriguing being by visiting anthropologists from India who studied, horrified, the way in which elderly people in the Netherlands live: isolated, institutionalized if in need of care, far from their families—and not even wishing their daughters to take care of them. Very Dutch, to be sure. But then again, this specific setup differs little from the arrangements in Germany, Sweden, Denmark—or a range of other European countries.

The boundaries around the Dutch state do not map on to a cultural domain. This is not to say that the cultural domain is simply larger; say *Europe*. There are striking differences between different European countries. For instance, Madeleine Akrich and Bernike Pasveer have com-

to our *quality of life* are "more than natural," but everything evoked in trials. The end points, the very goals of medical interventions, are essentially contested. They are intertwined with different, dissonant, ways of life. It is in this sense than one may say they are political. Take the question of how to compare bypass surgery and walking therapy for patients with atherosclerosis in their leg arteries. Which parameter should be improved after these treatments? If an angiographic picture were used to evaluate both treatments, walking therapy would never stand a chance of coming out as a successful intervention: it doesn't alter the width of the stenotic lumen. If pressure drop over the stenosis is measured, surgery will again appear to be the more successful treatment. As it even may if "a patient's pain-free walking distance after three weeks" is turned into the parameter of success. Walking therapy improves other parameters: it has different strengths. It is more likely to come out as a successful intervention if the patient's overall walking capabilities after six months are assessed. Or if the patient's gain in self-confidence is taken into account.

A politics-of-what assumes that the end points of trials, the goals sought for, are political in character. But there is more. Interventions have other effects, too. They bring about more than they seek to achieve. In current practice, trials deal with a few of these, so-called side effects. Usually, they take one or two calamities into account—like the risk of dying from an intervention (more real in bypass surgery than in walking therapy, although there, as everywhere in life, it may

pared childbirth practices in France and the Netherlands, countries only a few hours by car or train—but also worlds apart (Akrich and Pasveer 2001). Whereas in France pain is driven as much as possible from the scene of birth, Dutch women learn to *dive into* their pain, endure it, and use it to get attuned to—no, not just to what is happening to them, passively, but to what, actively, they are doing. Whereas in France women are tied to an apparatus measuring their physicalities, in the Netherlands they are advised to move about and find a position that best suits their bodies. Whereas a French father is just about allowed to be present during childbirth, a Dutch partner is expected to help his or her woman breathe properly to control her contractions. So there are

differences, contrasts. National cultures at last? No, Akrich and Pasveer hesitate to summarize their findings under two neat headings, French versus Dutch. What kind of entities are these? Where is the boundary between them? And what about the stories told to them by French women that resonate more with what happened in the Netherlands than with what went on in their own neighborhood—and vice versa?

So differences may be huge, even if not easy to *nationalize*. On the other hand, sometimes similarities between what is going on across borders are at least as impressive. But this raises further questions, instead of leading us to "cultural commonalties." Take David Armstrong's *Political Anatomy of the Body* (1983). To me, reading this book was astonishing. At the

occur). And alongside side effects, economic value enters evaluation: low cost is accepted as an important good. But most other aspects of the *mode of life* that come with enacting a disease in one way or another may enter clinical deliberation but have trouble getting represented in evaluation studies. Going twice daily for a serious walk requires a great amount of self-discipline: is that a good or not? Being taken care of by a devoted surgical team may be a treat—or not. And is it a rich experience or grossly alienating to become acutely aware of the color of the tissues beneath one's skin?

A politics-of-what explores the differences, not between doctors and patients, but between various enactments of a particular disease. This book has tried to argue that different enactments of a disease entail different ontologies. They each *do* the body differently. But they also come with different ways of *doing* the good. In each variant of atherosclerosis the *dis* of this *dis-ease* is slightly different. Different, too, are the ideals that, standing in for the unreachable "health," orient treatment. These and the other *goods* medicine tries to establish require further exploration. The study of ontology in medical practice presented here deserves to be followed up by an inquiry into the diverging and coexisting enactments of *the good*. Which goods are sought after, which bads fought? And in which ways are these goodnesses set up as being good—for there are huge

time, I was engaged in a joint research project into Dutch medical knowledge in (the second half of) the twentieth century. All *our* material was Dutch: medical professional journals, in Dutch, written by Dutch authors. And yet in Armstrong there were quotes that were almost literally the same. Armstrong attended to subtle shifts in professional investments in "the patient" and what this figure's characteristics implied for how patients should be listened to. This was one of our topics, too. And it became almost a game for me to compare the dates at which new configurations emerged. These did not run exactly parallel, but neither was one country always ahead of the other. Sometimes the British appeared to be a year or two earlier. At other times it was the Dutch. (If you want to make the comparison yourself, see Mol and Van Lieshout 1989.)

But what to make of this striking similarity? Instead of evoking *culture* in one way or another, it seems more promising to look at *money*. Due to the way health care was financed from the 1940s onward, general practitioners were relatively strong in Britain as well as in the Netherlands. In order to consolidate these arrangements, general practitioners came to stress their own specific strengths in contrast to those of the expanding medical specialists. These, they suggested, lay in the way they kept track of entire families over long periods of time, and not just individual patients; in attending not just to sick bodies but difficult life circumstances as well; and, finally, in conversational techniques that made them attentive to the points of view of their patients (techniques imported from the social workers with whom they collaborated in giving primary

differences between, say, conversational persuasion, scientific trials, ethical arguments, and economic power play. Or, another dimension of such a possible study, how do we live with lack and badness, and how do we practically handle the limits to the good?

These questions are not answered here. Investigating *the body multiple* merely helps to open them up. In suggesting that we pose them, however, there is a strong suggestion—or should I say sentiment?—that there is not such a thing as a single gradient of the good that "we" (whoever we are) might all agree about (whether convinced by the facts or in an open and honest discussion). Like ontology, the good is inevitably multiple: there is more than one of it. That is why for a politics-of-what the term *politics* is indeed appropriate. For a long time, and in many places, science held (or continues to hold) the promise of closure through fact-finding. In ethics, the promise of closure, or at least temporary consensus, through reasoning is widely shared. In an attempt to disrupt these promises, it may help to call "what to do?" a political question. The term *politics* resonates openness, indeterminacy. It helps to underline that the question "what to do" can be closed neither by facts nor arguments. That it will forever come with tensions—or doubt. In a political cosmology "what to do" is not given in the order of things, but needs to be established. Doing good does not follow on finding out about it, but is a matter of, indeed, doing. Of trying, tinkering, struggling, failing, and trying again.

care—who had in turn imported them from American social work and humanist psychology). And when they got a foothold in medical schools, general practitioners started to teach their conversational techniques to all future doctors. It is *this* that makes visiting a Dutch physician resemble visiting a British physician far more than, say, a German one—even if in seventy-five other ways the differences between Dutch and German "cultures" are much smaller.

So where a text comes from, how to specify its local provenance, is a topic rather than something to be taken for granted. This is an issue much discussed in recent anthropological literature, in part because delineating a site helps in its turn to specify what "a culture" is made to be.

(See, for example, the various texts assembled in Fog Olwin and Hastrup 1997.) Is the specificity of the material in this book its *Dutchness*? Is it that of a country with a generally well-educated population? Or does it have to do with a health care organization where general practitioners are relatively strong? Or with places where most patients get all their health care costs reimbursed? Or can this story only be understood as stemming from an academic hospital in a medium-sized town that is neither in the southern Catholic part of the Netherlands nor in the severely Protestant north? The possibilities are endless. They can be piled up to the point where the material analyzed here can be said to come from hospital Z and hospital Z alone.

Beyond Choice

The goodness ingrained in different versions of any one disease is inevitably contested. But this does not mean that a politics of what can depend on the traditional other of knowledge and reasoning: choice. Multiplicity, after all, is not quite pluralism. Diseases may be enacted differently at different sites, but the *sites* in question are not *sides*. Instead, different enactments of any one disease are interdependent. They may be added up; patients may be distributed between them; and they may include each other. There are innumerable tensions inside medicine but clashes between fully fledged paradigms are rare. Even an internist who scolds surgeons for focusing on clogged up lumens while not attending to the atherosclerotic process *has no choice* when called on by a patient whose ulcerating wounds no longer heal due to a lack of oxygen. He sends the patient (with collegial greetings) for an operation.

The interdependence between different versions of any one disease makes "choice" an ill-suited term for articulating the quintessence of a politics of what. And so does the interference between the enactment of disease(s) and that of other realities. Diseases, after all, are not the only phenomena enacted in the hospital. There are many more: sex difference, age and aging, Dutchness and foreignness, professionalism, emotional wisdom and instability, and so on. Thus, when two variants of a disease are separated out as each other's alternatives, a lot more is at stake than these variants alone. Take, for instance, the reality of "the sexes." It is implicated in enacting atherosclerosis. The more operations surgeons do, the more important the layers of fat underneath the human skin become for what it is *to be* or *not to be* a woman—or a man. But does

Let us now turn to sociology. Since hospital Z, and only hospital Z, is where the fieldwork for this study was done, many sociologists might be inclined to see it as a *micro*study. A study into something small. But is it? In his *Postmodern Geographies*, Edward Soja talks about Los Angeles (Soja 1989). A quite different place from hospital Z. Equally small? Well, measured in square kilometers it is somewhat larger, but those who set the *micro* against the *macro* might still say that it is pretty small. However, Soja escapes such attempts at scaling. He aptly shows how the town he chose to study includes "everything." It *all comes together in Los Angeles*—as one of his chapter headings goes. One reason for this is that people from literally all over the globe have come to live there. And they have brought their clothes, food, marriage customs, language—everything —with them. But Los Angeles is a big container for another reason. Everything that Soja takes to be crucially important for postmodern times can be signaled in this single city: all the shifts and changes that have to do with cities, their (absence of) planning, their distances, their patterns of trade, their transport systems—everything that geographers take to be important is

that argue for or against operations? With every health care statistic produced, the difference between the sexes gets more difficult to disentangle, for the pre-printed forms all ask one to come out as either M or F and thus tend to add yet another M/F difference to the pile that is already established. Setting normal values for each person individually, by contrast, empties out the relevance of differentiating between two sexed populations. It goes on and on. There are ever so many interferences between "atherosclerosis" and "sex difference." But how might those inform "choices" to be made as a part of a politics of what? It is difficult to see how to take just this single other relevance, let alone all realities enacted into account: one just cannot gain an overview. And one's evaluation of the enactment of any *one* object may well contradict one's evaluation of the other.

There is a third difficulty with the term "choice." If practices enact not just one entity, but evoke a world, then it is not only diseases that come in varieties, but people, too. They, or maybe it is better to say, *we,* whether figuring as professionals, patients, or something else, are caught up in this. We do not master realities enacted out there, but we are involved in them. There are, therefore, no independent actors standing outside reality, so to speak, who can choose for or against it. Take the surgeon. That figure varies along with the rest of reality. If atherosclerosis is enacted as a deviant condition that happens incidentally and accidentally, a surgeon is enacted as the welcome savior of an unfortunate patient. If, however, atherosclerosis is enacted as a slow process that should be

present in Los Angeles. And since it is all there, there is no need for the analyst to travel all over the place. There is no need to look for a *big* object. This single city serves. It contains everything.

The same is true for hospital Z. It makes little sense to talk of its size, let alone to call it small. Again, this is not just because the actual physical entities present in the building come from a lot of places. There are American journals; German measurement machines; Japanese televisions; computers made in the Philippines; there is South American coffee —and so on, as in all modern hospitals. People working in Z have also circulated—I mentioned before that some of them come from elsewhere (China, Portu-

gal, Switzerland, Britain) while others may have been born in the Netherlands but have spent a few years doing research in Paris, Seattle, or Toronto—or practicing in some small African town. But there is more. If one wanted to study, say, angiography, then what kind of large place would one try to find? Sure, there are hospitals slightly bigger than Z, but one cannot study the workings and usages of an X-ray apparatus somewhere "macro." It is always "micro," in a particular place. And the same is true for surgery: this is done on one body at a time. Or talking to a patient. Or thinking about how to treat. Adding up figures that come from ten or a hundred hospitals doesn't gives a *bigger* picture—it simply depicts something else. It

stopped early on, a surgeon is someone who is always too late—someone who only alleviates symptoms and is quite incapable of getting at the real disease. The identity of the very surgeons who might want to "choose" between modes of enacting atherosclerosis interferes with the "choices" to be made.

Patient identity is equally at stake. What a patient *is* is not given once and for all and is not so strongly established outside the hospital that it may be carried along into the consulting room, the ward, the operating theater, or the research lab. With every enactment of atherosclerosis there comes another patient. An example. If atherosclerosis is enacted as a genetically based deviance, you are simply burdened with it if you have the wrong genes. However, in so far as the development of the disease is enacted as a lifestyle matter, someone with atherosclerosis may be accused of having led a bad, unhealthy life. In this context, then, the patient is marked as irresponsible. This is not only a strange qualification to opt for if one had a choice, but also one that disqualifies one's abilities to handle options.

So there are incongruencies between what is implied in the notion of choice and the coexistences and interferences between different versions of reality described in this book. All in all, "choice" may not be the best term to capture what needs to be done, and what is going on, in the politics of what that *we* as medical professionals, ethnographers, sociologists, philosophers, and, yes, as patients too engage or may engage in. We need other terms. We have some other terms: discord, tension, contrast, multiplicity, interdependence, coexis-

conveys, say, epidemiological rather than individual facts; a numerical rather than a narrative reality; aggregations rather than events. (Why is it still necessary to argue against the idea that there is such a thing as a big picture? The argument was made *in the literature* quite a while ago. The scientist shuffling with paperwork on a desk, Bruno Latour explained in 1984 in French, handles not *more* variables than the hunter armed with arrows out in the field but, usually, far *less*. The scientist's numbers are just simplifications from a wide territory skillfully drawn together. And instead of residing in some *macro* place, they are on a desk. For the English version of this argument, see Latour 1988.)

Events are necessarily local. Somewhere. Situated. And in as far as this book tells about events, its object is necessarily local, too. But the main object of this book is not even events, but something different yet again. *Coexistence.* Theoretically speaking, this book is about the modes of coordination, distribution, and inclusion that allow different versions of a "single" object to coexist. But where, in what *place*, might coexistence be studied? There may be long distances between the entities that coexist under a single name. Take McDonald's. It is a fascinating multiple object, with endless similarities and differences between its various outlets, worldwide. (The idea that there is such a thing as the one and

tence, distribution, inclusion, enactment, practice, inquiry—but we could do with more. Which ones? That is another of the questions this book opens up rather than answers. For now, the point is this. In contrast with the universal-istic dreams that haunt the academic philosophical tradition, the world we live in is not one: there are a lot of ways to live. They come with different ontologies and different ways of grading the good. They are political in that the differences between them are of an irreducible kind. But they are not exclusive. And there is no *we* to stand outside or above them, able to master them or choose between them: we are implied. Action, like everything else, is enacted, too.

Clinical Medicine

That there are alternatives to each particular practice does not turn the hospi-tal, or health care, into a state of permanent turmoil. Tensions crystallize out into patterns of coexistence that tend to only gradually dissolve. Though noth-ing is sure or certain, the permanent possibility of doubt does not lead to an equally permanent threat of chaos. Even if stability is never reached, tensions are tamed. There are recurrent patterns of coexistence between different en-actments of any one disease. Addition, translation, distribution, inclusion: they keep the hospital together—just as they assemble the body and its diseases.

Describing health care in this way, or so I claim, is an act. How far this act may reach, whether and if so how this text will make a difference in practice remains to be seen. It depends on where this book is moved to, on who might run away with it, on the number of copies sold, on the (non)accidental overlaps between its concerns and those of some of its possible readers. What are you,

only, successfully globalized, McDonald's is done away with in Watson 1997.) But then again, if one is interested in modes of coexisting, it may well be that hospital Z *contains them all.* Coordination, distribu-tion, and inclusion, at least these three, are all to be found in Z. It isn't even neces-sary to roam through the entire building to achieve this—there are lots of sites and situations in hospital Z that aren't men-tioned in this book. A few practices relating to atherosclerosis of the leg vessels seem to compose a field *big enough* to contain as many patterns of coexistence as can be analyzed in a single book.

All this suggests that the precise *size* of a field is of little importance to the theo-rist who does not try to map that field, but tries to discern patterns in it, modes and modalities of, say, coexistence (but it might be something else as well). But if the size of the field is irrelevant—in-deed unmeasurable—this does not mean that the fact that there is a field is of no importance. The patterns of coexistence described here exist *somewhere.* Whatever the place is called: hospital Z; the enact-ments of atherosclerosis; health care; the Netherlands; the last decade of the twen-tieth century; well-insured surroundings;

reader, going to do with my words? That is beyond me—it is up to you. But I can try, have tried, to be articulate about what this text does intellectually. In theory, so to speak. It does not engage in criticism. I have *not* pointed, here, at the wrongs of medicine in general nor at those of the treatment of atherosclerosis in hospital Z in particular. I do not seek to confirm that all is well, but have argued, instead, that separating out right and wrong is only possible if one has a standard. I have not deployed such a standard here, but have analyzed the coexistences of different enactments of reality and have claimed that ever so many standards, different ways of grading *the good,* come with them.

However, this is not a neutral book either. Far from it. Analyzing medicine as enacting different realities and different ways of qualifying the good is not just a way of talking about medicine but also a way of talking inside it. Inside the medical world, this book is one of many voices that resist the idea that rationalization is the ultimate way of improving the quality of health care. Rationalization as an ideal starts from the idea that the problem with the quality of health care resides in the messiness of its practice. However, even if it may be messy, practice is something else as well: it is complex. The juxtaposition of different ways of working generates a complexity that rationalization cannot flatten out—and where it might, this is unlikely to be an improvement. In those sites and situations where a so-called scientific rationale (be it that of pathology, pathophysiology, or, most likely at the moment, that of clinical epidemiology) is brought into practice, with sufficient effort it may well come to dominate the other modes that are already at work. But this does not so much improve medicine as impoverish it. And that loss is borne by the clinic.

medical practice. There is a lot more to say about these ways of naming, localizing. But what I want to stress for now is just this one thing: that my theoretical investigation into the coexistence of the various versions of a multiple object were, indeed, *localized.* That a philosophical interest in ontology was linked up here with the empirical study of a field. This goes against the dominant tradition in philosophy. For a long time, the endeavors united under the banner of *philosophy* were presented as having a peculiar relation to place. They were *universal:* valid everywhere—and rooted nowhere in particular. Philosophical concepts had to

be of universal value. Norms had to be justified by arguments of universal pertinence. But all this could be done here and now. What was right in theory was supposed to be transportable anywhere—so easily that no attention was paid to what it might mean to transport "rightness." Universalities need no landing strips, telephone lines, or even satellites. The question of their transport is simply not posed. (For the obviously slightly more complicated history of the relation between philosophy and *place,* see Casey 1997.)

Some philosophers have opened up ways of leaving that dream of universality

In stressing multiplicity, this book lends support to clinical medicine. Clinical medicine is the tradition that departs from patient histories and presenting symptoms rather than from physicalities isolated in lab-like circumstances. The tradition, too, that lives with adaptable subjective evaluation rather than requiring objectified figures. A tradition of case histories rather than counting. This book doesn't support the clinical tradition by critically pointing out where it has lost, or is losing, ground. Instead, it does so by stressing its present, under-acknowledged, importance. The surgeons of hospital Z, after all, only open up arteries if their patients' daily lives are likely to gain from it as a result. Clinical considerations are crucial to their treatment decisions. And only those patients who present themselves with complaints make it to the hospital in the first place.

The proliferation of medical techniques may give reason to fear that the lab is taking over, but something quite different is equally possible. Since each diagnostic outcome diverges from the others, the idea of *gold standards* may get undermined rather than strengthened. And if each therapeutic intervention achieves something different, what counts as improvement may similarly tend to become less obvious. The question "is this intervention effective" then dissolves into another question: "what effects does it have?" Clinical considerations, however fuzzy they may be, however badly they fit into forms and accounting systems, may well prove obdurate and tenacious. After all, they con-

behind. Walter Benjamin offers a wonderfully radical example. His *Passagen-Werk* (1999) was *both* situated *in* philosophy *and* somewhere *earthly* in particular. Paris. The modern city. Its architecture. Arcades. Encounters between strangers. It is this overt attentiveness to the *situatedness* of thinking (its objects, its possibilities, its enactment, its preformative effects) that marks the philosophical literatures that figure in the background of this book and form its venerated ancestors. The one to conclude with is Michel Foucault. In his writings it is an acute sense of situatedness that turns philosophy into something worthwhile in the first place. Something forever shifting, changing. A mode of engaging in philosophy that advertises itself as linked up with the here-and-now, with *ourselves,* cannot be—nor does it hope to be—*universal.* It is localized. Foucault mainly explored empirical matters in a historical manner—but ethnographic or rather praxiographic extensions easily follow. Please add, therefore "topographical" to the "historical" situatedness that figures in the following quote. So that this subtext relating to the literature may end, as is only fitting, with words *taken from* the literature. "The critical ontology of ourselves has to be considered not, certainly, as a theory, a doctrine, nor even as a permanent body of knowledge that is accumulating; it has to be conceived as an attitude, an ethos, a philosophical life in which the critique of what we are, is at one and the same time the historical analysis of the limits that are imposed on us and an experiment with the possibility of going beyond them" (Foucault 1984, 50).

cern daily lives. And daily life is what, when it comes to it, matters most to people. It is where patients, *we,* have to live with our doubts and our diseases. No, all is *not* well. But where rationalization risks to overrule the clinical tradition with ever more statistics, accounting systems, figures, and other carriers of scientificity, this book sides with those voices that seek to improve the clinic on its own terms. Which terms? How to *do* the clinical good better? These are further questions I leave open here.

So even if it is not critical, this is not a neutral study. There are other modes of partiality than that of passing judgment. Undermining the traditional hierarchy between the sciences is a way of strengthening the disciplines that occupy the lower ranks in the hierarchy. Pointing at the persistent possibility of doubt eats at the self-assuredness (and the convincing power) of the techniques that claim that they are finally able to bring light and science to messy practices. Rather than comparing different interventions within a given dimension, laying open the various dimensions of comparability makes space for and gives visibility to dimensions that currently attract the least attention. Not going primarily with a politics of *who* but stressing the necessity of a politics of *what* helps to open up the professional domain instead of pushing it back. And doubting whether choice is the best term to use in a politics of what (a politics that includes ontology rather than presuming it) acts against rationalist fantasies of what it is to strive after the good. Presenting the *body multiple* as the reality we live with is not a solution to a problem but a way of changing a host of intellectual reflexes. This study does not try to chase away doubt but seeks instead to raise it. Without a final conclusion one may still be partial: open endings do not imply immobilization.

bibliography

Akrich, M., and D. Pasveer. 2001. Obstetrical trajectories. On training women/bodies for (home) birth. In *Birth by design*. Edited by R. DeVries, C. Benoit, E. van Teijlingen, and S. Wrede. London: Routledge.

Andersen, T. F., and G. Mooney, eds. 1990. *The challenges of medical practice variations*. Houndmills, UK: Macmillan.

Anselle, J.-L. 1990. *Logiques métisses: Anthropologie de 1'identité en Afrique et ailleurs*. Paris: Payot.

Armstrong, D. 1983. *Political anatomy of the body: Medical knowledge in Britain in the twentieth century*. Cambridge: Cambridge University Press.

———. 1988. Space and time in British general practice. In *Biomedicine examined*. Edited by Margaret Lock and Deborah Gordon. Dordrecht: Kluwer.

Arney, W., and B. Bergen. 1984. *Medicine and the management of living: Taming the last great beast*. Chicago: University of Chicago Press.

Ashmore, M. 1989. *The reflexive thesis: Wrighting sociology of scientific knowledge*. Chicago: University of Chicago Press.

Ashmore, M., M. Mulkay, and T. Pinch. 1989. *Health and efficiency: A sociology of health economics*. Milton Keynes: Open University Press.

Barker, M. 1982. *The new racism*. London: Junction Books.

Barreau, H., ed. 1986. *Le même et 1'autre: Recherches sur 1'individualité dans les sciences de la vie*. Paris: Éditions du CNRS.

Bazerman, C. 1988. *Shaping written knowledge*. Madison: University of Wisconsin Press.

Benedict, R. 1997. A note on Dutch behaviour. In *Notities over Nederlanders*. Edited by R. van Ginkel. Amsterdam: Boom.

Benhabib, S., ed. 1996. *Democracy and difference: Contesting the boundaries of the political*. Princeton: Princeton University Press.

Benjamin, W. 1999. *The Aracdes project*. Translated by Howard Eiland and Kevin McLaughlin. Cambridge: Harvard University Press.

Berg, M. 1997. *Rationalizing medicine: Decision-support techniques and medical practices*. Cambridge, Mass.: MIT Press.

Böhme, G. 1980. *Alternative der Wissenschaft*. Frankfurt: Suhrkamp.

Butler, J. 1990. *Gender trouble: Feminism and the subversion of identity*. London: Routledge.

Canguilhem, G. [1966] 1991. *The normal and the pathological*. New York: Zone Books.

Caplan, A. L., H. T. Engelhardt, Jr., and J. J. McCartney. 1981. *Concepts of health and disease: Interdisciplinary perspectives*. Reading, Mass.: Addison-Wesley.

Casey, E. 1997. *The fate of place: A philosophical history*. Berkeley: University of California Press.

Chauvenet, A. 1978. *Médecine au choix, médecine de classes*. Paris: PUF.

Clifford, J. 1988. *The predicament of culture: Twentieth-century ethnography, literature, and art*. Cambridge, Mass.: Harvard University Press.

Clifford, J., and G. E. Marcus, eds. 1986. *Writing culture*. Berkeley: University of California Press.

Cussins, C. 1996. Ontological choreography. *Social Studies of Science* 26:575–610.

De Laet, M., and A. Mol. 2000. The Zimbabwe bush pump: Mechanics of a fluid technology. *Social Studies of Science* 30:225–63.

Dehue, T. 1995. *Changing the rules*. Cambridge: Cambridge University Press.

Dodier, N. 1993. *L'expertise médicale*. Paris: Métailié.

———. 1994. Expert medical decisions in occupational medicine: A sociological analysis of medical judgement. *Sociology of Health and Illness* 16(4):489–514.

Doyal, L., and I. Pennel. 1979. *The political economy of health*. London: Pluto Press.

Dreifus, C., ed. 1978. *Seizing our bodies: The politics of women's health*. New York: Vintage House.

Duden, B. 1991. *The women beneath the skin: A doctor's patients in eighteenth-century Germany*. Translated by Thomas Dunlop. Cambridge, Mass.: Harvard University Press.

Duyvendak, J. W., ed. 1994. *De verzuiling van de homobeweging*. Amsterdam: Sua.

Engel, G. 1981. The need for a new medical model. In *Concepts of health and disease*. Edited by A. Caplan, H. T. Engelhardt, Jr., and J. McCartney. Reading, Mass.: Addison-Wesley.

Engelhardt, H. T., Jr. 1975. The concepts of health and disease. In *Evaluation and explanation in the biomedical sciences*. Edited by H. T. Engelhardt, Jr. and S. Spicker. Dordrecht: Reidel.

Engelhardt, H. T., Jr., and A. Caplan. 1987. *Scientific controversies*. Cambridge: Cambridge University Press.

Epstein, S. 1996. *Impure science: AIDS, activism, and the politics of knowledge*. Berkeley: University of California Press.

Fleck, L. [1935] 1980. *Entstehung und Entwicklung einer wissenschaftlichen Tatsache.* Frankfurt: Suhrkamp.

Fog Olwin, K., and K. Hastrup, eds. 1997. *Siting culture: The shifting anthropological object.* London: Routledge.

Foucault, M. 1973. *The birth of the clinic: An archaeology of medical perception.* London: Tavistock.

———. 1979. *Discipline and punish.* Translated by Alan Sheridan. New York: Vintage.

———. 1981. *The history of sexuality, vol I: An introduction.* Harmondsworth, U.K.: Penguin.

———. 1984. What is enlightenment? In *The Foucault Reader,* Edited by P. Rabinow. New York: Pantheon.

Fox, N. 1994. Anaesthetists, the discourse on patient fitness, and the organisation of surgery. *Sociology of Health and Illness* 16:1–18.

Freidson, E. 1970. *The profession of medicine.* New York: Harper and Row.

Gilman, S. 1985. *Difference and pathology: Stereotypes of sexuality, race, and madness.* Ithaca: Cornell University Press.

Ginkel, R., van. 1997. *Notities over Nederlanders.* Amsterdam: Boom.

Goffman, E. [1959] 1971. *The social presentation of self in everyday life.* Harmondsworth, U.K.: Penguin.

Goodman, N. 1978. *Ways of worldmaking.* Indianapolis: Hackett Publishing Company.

Hacking, I. 1992. The self-vindication of the laboratory sciences. In *Science as practice and culture.* Edited by A. Pickering. Chicago: University of Chicago Press.

Hahn, R. 1985. A world of internal medicine: Portrait of an internist. In *Physicians of Western Medicine.* Edited by R. Hahn and A. Gaines. Dordrecht: Reidel.

Haraway, D. 1989. *Primate visions.* New York: Routledge.

———. 1991. Gender for a Marxist dictionary: The sexual politics of a word. In *Simians, cyborgs, and women: The reinvention of nature.* London: Free Association Press.

———. 1997. *Modest_Witness@Second_Millennium.FemaleMan© Meets_OncoMouse™.* New York: Routledge.

Harding, S. 1986. *The science question in feminism.* Ithaca: Cornell University Press.

Harvey, D. 1990. *The condition of postmodernity.* Oxford: Basil Blackwell.

Helman, Cecil. 1988. Psyche, soma, and society: The social construction of psychosomatic disorders. In *Biomedicine examined.* Edited by Margaret Lock and Deborah Gordon. Dordrecht: Kluwer.

Henderson, L. J. 1935. Physician and patient as a social system. *New England Journal of Medicine* 212:819–23.

Hirschauer, S. 1993. *Die soziale Konstruktion der Transsexualität: Über die Medizin und den Geslechtswechsel.* Frankfurt: Suhrkamp.

Jacobus, M., E. Fox Keller, S. Shuttleworth, eds. 1990. *Body/politics: Women and the discourses of science.* New York: Routledge.

Kondo, D. 1990. *Crafting selves: Power, gender, and discourses of identity in a Japanese workplace*. Chicago: University of Chicago Press.

Kuhn, T. 1962. *The structure of scientific revolutions*. Chicago: University of Chicago Press.

Lacoste, Y. 1976. *La géography, ca sert, d'abord, à faire la guerre*. Paris: Maspero.

Lakatos, I., and A. Musgrave. 1970. *Criticism and the growth of knowledge*. Cambridge: Cambridge University Press.

Lakoff, G., and M. Johnson. 1979. *Metaphors we live by*. Chicago: University of Chicago Press.

Latour, B. 1987. *Science in action*. Milton Keynes: Open University Press

———. 1988. *The pasteurization of France*. Cambridge, Mass.: Harvard University Press.

———. 1993. *We have never been modern*. New York: Harvester Weathsheaf.

———. 1996. *Aramis: Or the love of technology*. Cambridge, Mass.: Harvard University Press.

Latour, B., and S. Woolgar. 1979. *Laboratory life*. London: Sage.

Law, J. 1994. *Organizing modernity*. Oxford: Blackwell.

———. 2000. On the subject of the object: Narrative, technology, and interpellation. *Configurations* 8:1–29.

———. 2002. *Aircraft stories. Decentering the objects in technoscience*. Durham: Duke University Press.

Law, J., and A. Mol. 1995. Notes on materiality and sociality. *Sociological Review* 43:274–94.

Lecourt, D. 1976. *Lyssenko*. Paris: Maspero.

Lijphart, A. 1968. *The politics of accommodation: Pluralism and democracy in the Netherlands*. Berkeley: University of California Press.

Lynch, M., and S. Woolgar, eds. 1990. *Representation in scientific practice*. Cambridge, Mass.: MIT Press.

MacKenzie, C., and B. Barnes. 1979. Scientific judgement: The biometry-mendelism controversy. In *Natural order*. Edited by B. Barnes and S. Shapin. London: Sage.

McCrea, F., and G. Markle. 1984. The estrogen replacement controversy in the USA and the UK: Different answers to the same question? *Social Studies of Science* 14:1–26.

Mol, A., and J. Law. 1994. Regions, networks, and fluids: Anaemia and social topology. *Social Studies of Science* 24:641–71.

Mol, A., and M. Berg. 1998. Introduction. In *Differences in medicine: Unravelling practices, techniques, and bodies*. Edited by Marc Berg and Annemarie Mol. Durham: Duke University Press.

Mol, A., and P. van Lieshout. 1989. *Ziek is het woord niet: Medicalisering en normalisering in de Nederlandse huisartsgeneeskunde en geestelijke gezondheidszorg, 1945–1985*. Nijmegen: SUN.

Moore, B. 1966. *Social origins of dictatorship and democracy*. Harmondsworth, U.K.: Penguin.

Mouffe, C. 1993. *The return of the political*. London: Verso.

Okeley, J., and H. Callaway, eds. 1992. *Anthropology and autobiography*. London: Routledge.

Parsons, T. 1951. *The social system*. New York: Free Press.

Pickering, A., ed. 1991. *Science as practice and culture*. Chicago: University of Chicago Press.

Pool, R. 1989. Gesprekken over ziekte in een Kameroenees dorp: Een kritische reflectie op medisch-antropologisch onderzoek. In *Ziekte, gezondheidszorg, en cultuur*. Edited by S. van der Geest and G. Nijhof. Amsterdam: Het Spinhuis.

Pool, R. 1994. *Dialogue and the interpretation of illness: Conversations in a Cameroon village*. Oxford: Berg Publishers.

Poulanzas, N. 1978. *L'État, le pouvoir, le socialisme*. Paris: PUF.

Rose, S., ed. 1982. *Against biological determinism*. London: Allison and Busby.

Serres, M. 1980. *Le passage du nord-ouest*. Paris: Éditions de Minuit.

Serres, M. 1994. *Atlas*. Paris: Julliard.

Shapin, S., and S. Schaffer. 1985. *Leviathan and the air-pump: Hobbes, Boyle, and the experimental life*. Princeton: Princeton University Press.

Showalter, E. 1985. *The female malady: Women, madness, and English culture, 1830–1980*. London: Virago Press.

Soja, E. 1989. *Postmodern geographies: The reassertion of space in critical theory*. London: Verso.

Star, S. L., and J. Giesemer. 1989. Institutional ecology, translations, and boundary objects: Amateurs and professionals in Berkeley's Museum of Vertebrate Zoology, 1907–1939. *Social Studies of Science* 19:387–420.

Stepan, N. 1987. Race and gender: The role of analogy in science. *Isis* 77:261–77.

Stocking, G. 1968. *Race, culture, and evolution*. New York: Free Press.

Strathern, M. 1991. *Partial connections*. Savage, Md.: Rowman and Littlefield.

———. 1992a. *After nature: English kinship in the late twentieth century*. Cambridge: Cambridge University Press.

———. 1992b. *Reproducing the future: Anthropology, kinship, and the new reproductive technologies*. Manchester: Manchester University Press.

Strauss, A. 1978. *Negotiations*. San Francisco: Jossey-Bass.

Suppe, F., ed. 1977. *The structure of scientific theories*. Urbana: University of Illinois Press.

Sullivan, Mark. 1986. In what sense is contemporary medicine dualistic? *Culture, Medicine, and Psychiatry* 10:331–50.

Trinh Minh-ha. 1989. *Women native other*. Bloomington: Indiana University Press.

Watson, J., ed. 1997. *Golden arches east: McDonalds in East Asia*. Stanford: Stanford University Press.

Willems, D. 1992. Susan's breathlessness: The construction of professionals and laypersons. In *The social construction of illness*. Edited by Jens Lachmund and Gunnar Stollberg. Stuttgart: Franz Steiner Verlag.

Young, A. 1981. When rational men fall sick: An inquiry into some assumptions made by medical anthropologists. *Culture, Medicine, and Psychiatry* 5:317–35.

Yoxen, Edward. 1982. Constructing genetic diseases. In *The problem of medical knowledge: Examining the social construction of medicine.* Edited by P. Wright and A. Treacher. Edinburgh: Edinburgh University Press.

Page numbers in italics refer to subtext.

Addition, 7, 68–72
Admission: to hospital, 55, 74, 128–30
Akrich, M., *174–75*
Amputation, 18, 44, 127
Anatomy, 48, 109, 162
Andersen, T., *2*
Angiography, 72–83, 93–94, 146
Angiology, 40–41, 95–102
Ankle/arm index, 60–70, 100
Anselle, J.-L., *132–33*
Anthropology: medical, *14–17, 23–24*
Armstrong, D., *128*, *175–76*
Arney, W., *128*
Ashmore, M., *155*, *161–64*
Association, *61–65*
Atherosclerosis: lumen loss, 74–83; small lumen, 29–38, 46, 60. *See also* Intermittent claudication; Pain: on walking; Stenosis

Barker, M., *18–22*
Barnes, B., *91*
Barreau, H., *129*

Bazerman, G., *159*
Behabib, S., *106*
Benedict, R., *172–73*
Benjamin, W., *183*
Berg, M., *4*, *165–66*
Bergen, B., *128*
Blood, 73, 98, 125; clotting mechanism of, 108–14; flow, 74–83; platelets, 109; pressure, 58–66, 90; velocity, 75
Body: coherent, 55–72; as model of society, *54–84*; multiple, 7, 55, 177; virtual, 119
Böhme, G., *153*
Boundary, *135–42*; between disciplines, *17–22*; national, *171–75*; object, 138
Butler, J., *37–39*

Callaway, H., *159*
Callon, M., *34*
Cancer, 41, 169, 174
Canguilhem, G., *54–56, 121–24*
Caplan, A., *93–99, 121*

Casey, E., *182*
Change, 113–17
Chauvenet, A., *5*
Choice, 178–84
Cholesterol, 131–32, 138–42. *See also* Lipoproteins
Citizen, 167–72
Clash, 6, 62–66, 82, 97, 103, 108, 139
Clifford, J., *77–78, 159*
Clinical diagnosis, 51, 58, 60–66, *123–27, 141–42*
Clinical medicine, 181–84
Clinical trial, 67, 99–101, 140–42, *166–70, 172–82*
Closure: of controversy, 94–105, *102–14, 177*
Coherence, 55–66, *64–68, 95–96*
Coincidence, 45, 61; absence of, 68
Complaints, 41–51, 135; relation to laboratory findings, 60–66
Condition: versus process, 102–17
Construction, 32, 41–42
Controversy, 62, 87, 88–100, 99, 104–15
Coordination, 53–85
Corpse, 37, 45–46, 49–50, 125–26
Correlation, 76–83
Costs: monetary, 47, 95, 137, 147, 170
Criticism, 10, 47, 103, 154
Culture, *77–82, 177. See also* Nature
Cussins, C., *43*
Customer, 167–72
Cyborg, 137, *149*

Death, 55, 98, 128–33, 137–41. *See also* Corpse
Decision, 93–95, 168–82; -making meeting, 70–71, 82, 142, 146
Dehue, T., *153*
De Laet, M., *142*
Diabetes, 8, 63, 65, 105–6, 127, 134
Diagnosis, 22–26, 39–43

Difference, 38, 55, 62, *73–76*, 82, *114–16, 120–42, 174–82. See also* Multiplicity
Discourse, *61, 69*
Disease, 60–61, 84, 127, 156, 163; versus illness, 7–27, *10–12*
Distribution, 96–99, 99, 104, *107–8*, 108, *115–77*
Dodier, N., *68, 112–13*
Doppler measurement, 58–66, 130
Doubt, 47, 100, 161–66, *163*, 177, 184
Drug, 112–14
Duden, B., *25–26*
Duplex, 72–83
Duyvendak, J.-W., *107*

Effects: material, 89; of treatment, 66–67, *183*
Empirical philosophy, 4–7
Enactment, 32–36, 40, 44, 46–48, 54, 91–96, 154–60
Endarterectomy, 90–91, 100–102
Engelhard, T., *93–99, 110–11, 121*
Epidemiology, 128–33; clinical, 140–42, *168*
Epistemology, 5–7, 45–50, 149–66
Epstein, S., *167–70*
Ethnography, 7–27; of disease, 151–57; of knowledge practices, 5–7. *See also* Praxiography
Event, 15–27
Exception, 49–51, 63, 83
Explanation, 46–48, 60–69, 82, *103*

Fat: balances in blood, 103–7; subcutaneous, 90, 123, 147–48
Feelings, 8–9, 14, *25–26, 61–63*
File, 24, 48–49, 89
Fleck, L., *129–31*
Flow: chamber, 109–14. *See also* Blood flow
Fluid, *147*
Fog Olwin, K., *177*

Foucault, M., *46–47, 56–62, 66–70,*
125–28, 183
Foundation, 36–42, 45–50, 165
Fox, N., 111–12
Fox Keller, E., *187*
Fragmentation, 55, 72, 116
Freidson, E., *4–6*
Functionalism, *7–13*

Gender, *20–22,* 145. *See also* Sex-
difference
General practitioner, 22, 26, 63, 95,
130–31, 135–36, 139, *175–77*
Giesemer, J., *138*
Gilman, S., *124*
Ginkel, R. van, *172–74*
Gofmann, E., *34–37, 54*
Good, 165–66, 176–77
Goodman, N., *67*

Hacking, I., *74–76*
Hahn, R., *15*
Haraway, D., *19–22, 136–38, 149, 155–
57*
Harding, S., *154*
Harvey, D., *139*
Hastrup, K., *177*
Helman, C., *42*
Hematologist, 108–14
Henderson, L., *8*
Hierarchy, 62–64, 82
Hirschauer, S., *38–41*

Identity, *34–44,* 82, 180
Illness, 7–27, 71. *See also* Disease
Improvement, 134–41
Inclusion, 130–33, 143–49
Incommensurability, 75, 85
Incompatibility, 35, 88, 121
Indication criteria, 101–2
Individual: versus population, 127–42
Interaction: doctor-doctor, 26, 44,
64–66, 70–73, 82–83, 90, 98–100,

103–4, 111; doctor-patient, 21–25,
41–42, 69, 135
Interference, 121, 142–49, 152
Intermittent claudication, 22–23, 25,
114. *See also* Pain: on walking
Internist, 49, 103–8, 131
Interobserver variability, 74, 78
Intervention: reason for versus target of,
94; theory as, 165; therapeutic, 175
Interview: clinical, 22–28, 34, 40–43,
65; ethnographic, 8–13, 78, 131–33,
141
Intransitivity, 121–33, 145–49

Jacobus, M., *152*
Johnson, M., *91*

Knowledge: embedded, 15–17, 31; incorporated, 48–50; versus intervention,
89; material, 50, 62; valid, 152. *See
also* Epistemology
Kondo, D., *133–35*
Kuhn, T., *71–77*

Laboratory: and clinics, 66–72, 123,
141–43; research, 108–12
Lacoste, Y., *139*
Lakatos, I., *88*
Lakoff, G., *90*
Latour, B., *30–33, 34, 42, 62–66, 108–9,
140, 160, 180*
Law, J., *42–43, 68–69, 142, 159*
Lecourt, D., *99*
Lieshout, P. van, *128, 176*
Life: daily, 14–20, 70, 184; quality of,
174–82
Lijphart, A., *106–8*
Lipoproteins, 104–7
Literature: relating to, *2–30, 41–43*
Lyssenko affair, *98–99*

MacKenzie, D., *91*
Marcus, G., *159*

Markle, G., *101–5*

McCartney, J., *121*

McCrea, F., *101–5*

Meaning, 8–12, 75, 173

Method: ethnographic/praxiographic, 1–27, 31–33, *44, 34–50*, 48–51, 91, 99, 110, 123, 125, 127, 146, *152–60*

Microscope, 29–33, 38

Modes of ordering, *68–69*

Mol, A., *4, 42, 128, 142, 176*

Mooney, G., *2*

Moore, B., *106*

Motivation, 69–71

Mouffe, C., *114–15*

Mulkay, M., *161–64*

Multiplicity, 5, 50, 65–85, *83–85*, 91, 117, 143, 151–60, 164, 183

Musgrave, A., *88*

Nature: versus culture, *17–22, 33–44*

Netherlands, the, 1, 41, 49–50, 63, 128–32; 'Dutchness,' 24, 50, *170–77*, 178

Network, 65, *69–71*, 139–42

Normality: in contrast with pathological, *56–61*, 120–28

Normativity, 55–61, 121, 172–82; epistemological, *98–99*; ethnographic, 157–60

Numbers, 63–66, 105, 128–33

Nurse, 90–91, 105, 124, 173

Object, *76*; classes or versions, 159; common, 84; composite, 71; singular, 62–66; versus subject, *30–50*; virtual, 163. *See also* Ontology

Okeley, J., *159*

Ontology, 6–7, 83–86, 115, 149–50, *183*; ontological pyramid, 120, 153–60

Operation, 17–19, 83, 90–93, 97–102, 134–41, 147–48

Outpatient clinic, 21–26; versus pathology, 33–48

Pain, 20, 34–45, 58–66, 97–102; on walking, 22–23, 42

Paradigm, 71–76, 84

Parameter, 67, *168*, 141, 173–76

Parsons, T., *7–13, 23, 54, 57–58*

Partial connection, *16, 80–82*

Pasveer, B., *174–75*

Pathologist, 29–51

Pathology, 29–51, *45–50*, 125–26. *See also* Normality

Patient: as ethnographer, 13–19; as having a perspective, 7–13; as object of treatment, 122–26; presenting information, 41–42; as representing an interest, 168–72. *See also* Feelings; Motivation; Population

Peak systolic velocity, 76–83

Percutaneous transluminal angioplasty (PTA), 67, 96–102

Performance, 32–33, *34–44*

Perspective, 7–13, 67, 152

Philosophy, *30–32*, 152–66, 181; of science, 88–100

Physical examination, 51–52

Pinch, T., *161–64*

Politics, 95–100, *105–8, 114–16*; of what, 172–81; of whom, 166–72; of writing, 159–60

Pols, Jeannette: as interviewer, 13–19

Pool, R., *23–24*

Population, 47, 127–42, 130, 147. *See also* Clinical trial

Poulanzas, N., *136*

Practicalities, 15, 31, 36, 66, 119, 156

Practice, 5–6, 13, *30–33, 30–36*, 54, 91, 95–96, *108–11*, 110–14, 129–33, 140, 164

Pragmatism, 6, 43–48, 102, 109–16

Praxiography, 31–33, 53–55, 119–21, 142–49, 156–60; praxiographic turn, 83, 150. *See also* Ethnography

Process: versus condition, 102–17

Professionals, *5–6, 10–13*, 27, 57, *64–65*, 166–72
Psychosocial matters, 12, 27, 44–50, *133–34*, 153–55
Pulsations: of the arteries, 25, 33–34, 38, 51

Radiologist, 72–83, 93–102
Rationality, *14–15, 93–100*
Rationalization, *161–70*, 184
Reality: day-to-day, 27; enacted, 47–48; hidden, 61, 163; by itself, 9, 36; lived, 68; partaking in, 154; physical, 20, 125; in practice, 13, 164; social, 70–72; in treatment, 102; underlying, 36, 116; visible, 83; versus what seems-to-be, 82
Representation, 54, 70, 82, 90
Research, 75–82, 103–15, 161; publications, 87
Risk: of death, 101, 132; of developing atherosclerosis, 114, 137, 141; of side effects, 47, 175
Rose, S., *18*

Schaffer, S., *153*
Science, *71–76, 88–104, 115–16, 130, 136*; social studies of, 87, *88–104*
Self: versus other, *129–35*
Serres, M., *144–47*
Sex-difference, *37–40, 129–33, 144–49, 178–79*
Shapin, S., *153*
Showalter, E., *124*
Shuttleworth, S., *152*
Singularity, 55–65, 119, 163–64
Situatedness, 48–51, 55; of present text, *170–83*
Social: order, 54–61; reality, 70–72; science, 53; system, 7–13, 54–56; worlds, *67*
Society, *54–84, 136*
Sociology: medical, *3–13*

Soja, E., *178–79*
Sound, 58–62, 75, 125
Space, *139, 142*; non-Euclidian, 119, *144*
Star, L., *138*
Stenosis, 73–83, 93–96
Stepan, N., *124*
Stocking, G., *131*
Strathern, M., *18–23, 78–82, 147–49*
Strauss, A., 67
Subject: as human versus natural, *33–44*; as knower versus known, *44–49*; versus object, 31–33
Sullivan, M., *44–48*
Suppe, F., *152*
Surgeon, 21–27, 42, 51, 69–85, 90–102, 123–24, 142–43, 147–48, 179

Table: autopsy, 125; examination, 24, 72–75; operating, 19
Technician, 31, 39, 58–66, 74–81, 125–26
Technique, 33; diagnostic, 63–68, 72–83; operation, 90–91; safety of, 74–75. *See also* Risk
Tension, 61–66, 88–96, 102–7, *113–16*, 134–41
Thompson, C. *See* Cussins, C.
Topography, 37, 48–51
Touch, 12, 25, 33, 91, *158*
Transitivity. *See* Intransitivity
Translation, 35, 74, 78–85
Treatment: design of, 69–83; invasive, 90–102. *See also* Effects
Trial. *See* Clinical trial
Trinh, M., *159*
Truth, 7, 75, 182

Ultrasound, 58–66, 74–75. *See also* Doppler measurement; Duplex
Universality, 54, 114, 181, *182*

Visibility, 22–24, 30–32, 38–40, 73, 78–83, 89, 96–102, 123, 162

Walking therapy, 95–97, 162. *See also* Pain: on walking

Watson, J., *181*

West, the: medicine in, *2–3, 24;* mortality, 113; population of, 131; tradition, *135–36;* Western countries, 49–50

Whole, *73,* 77–84, 123, 130, 132

Willems, D., *146–47*

Woolgar, S., *42*

Writing. *See* Politics: of writing

X ray: machine, *72–73;* protection against, *72–73. See also* Angiography

Young, A., *14–17*

Yoxen, E., *42*

Annemarie Mol is Socrates Professor of Political Theory at the
University of Twente, Enschede, The Netherlands. She has published
widely in Dutch and English; is co-author of a book in Dutch on
medical discourse; editor, with John Law, of *Complexities: Social
Studies of Knowledge Practices* (Duke University Press 2002); and
editor, with Marc Berg, of *Differences in Medicine: Unraveling
Practices, Techniques, and Bodies* (Duke University Press 1998).

Library of Congress Cataloging-in-Publication Data

Mol, Annemarie.

The body multiple : ontology in medical practice / Annemarie Mol.

p. cm. — (Science and cultural theory)

Includes bibliographical references and index.

ISBN 0-8223-2902-6 (cloth : alk. paper)

ISBN 0-8223-2917-4 (pbk : alk. paper)

1. Atherosclerosis—Netherlands.   2. Social medicine.

3. Medical anthropology.   4. Ontology.   5. Medicine—Philosophy.

6. Ethnographic informants—Netherlands.   I. Title.   II. Series.

RC692 .M59   2002

362.1'96136—dc21   2002006932